Insulin

History of Health and Illness Series

Stuart Bradwel, *Insulin: A Hundred-Year History*
Jonathan Sadowsky, *The Empire of Depression: A New History*
Orna Ophir, *Schizophrenia: An Unfinished History*

Insulin

A Hundred-Year History

Stuart Bradwel

polity

The right of Stuart Bradwel to be identified as Author of this Work has been asserted in accordance with the UK Copyright, Designs and Patents Act 1988.

First published in 2023 by Polity Press

Polity Press
65 Bridge Street
Cambridge CB2 1UR, UK

Polity Press
111 River Street
Hoboken, NJ 07030, USA

ISBN-13: 978-1-5095-5072-2

A catalogue record for this book is available from the British Library.

Library of Congress Control Number: 2022949655

Typeset in 10.75 on 14 Adobe Janson
by Cheshire Typesetting Ltd, Cuddington, Cheshire
Printed and bound in Great Britain by CPI Group (UK) Ltd, Croydon

The publisher has used its best endeavours to ensure that the URLs for external websites referred to in this book are correct and active at the time of going to press. However, the publisher has no responsibility for the websites and can make no guarantee that a site will remain live or that the content is or will remain appropriate.

Every effort has been made to trace all copyright holders, but if any have been overlooked the publisher will be pleased to include any necessary credits in any subsequent reprint or edition.

For further information on Polity, visit our website:
politybooks.com

Contents

Abbreviations

AAP	Association of American Physicians
ACA	Affordable Care Act
ADA	American Diabetes Association
BDA	British Diabetic Association (Operating as Diabetes UK from 2000)
BERTIE	Beta Cell Education Resources for Training in Insulin and Eating
BMJ	British Medical Journal
CGM	continuous glucose monitoring
CSII	continuous subcutaneous insulin infusion
DAFNE	Dose Adjustment for Normal Eating
DCCT	Diabetes Control and Complications Trial
DESG	Diabetes Education Study Group
DKA	diabetic ketoacidosis
DTTP	Diabetes Teaching and Training Programme
EASD	European Association for the Study of Diabetes
EPP	Expert Patients Programme
FIT	functional insulin therapy
GRI	glucose responsive insulin
HbA1c	glycated haemoglobin
HHS	hyperosmolar hyperglycaemic state

ICT	insulin coma therapy
IDF	International Diabetes Federation
IQWiG	Institute for Quality and Efficiency in Healthcare (Germany)
JDRF	Juvenile Diabetes Research Foundation
MDI	multiple daily injections
MRC	Medical Research Council (UK)
NDDG	National Diabetes Data Group (USA)
NHS	National Health Service (UK)
NPH	Neutral Protamine Hagedorn [Insulin]
OIF	Open Insulin Foundation
PAP	Patient Assistance Programme
PZI	Protamine Zinc Insulin
SBGM	self blood glucose monitoring
T1DM	Type 1 diabetes mellitus
T2DM	Type 2 diabetes mellitus
TGH	Toronto General Hospital
UKPDS	UK Prospective Diabetes Study
WHO	World Health Organization

Figures

Acknowledgements

No book is the product of one person alone. Without the help and generosity of others, I would almost certainly have never finished writing this one. Perhaps most importantly, I would like to express my eternal gratitude to those who agreed to participate in interviews – people with diabetes and healthcare professionals alike – while I was completing my PhD studies at the University of Strathclyde. Many of the testimonies that resulted are discussed throughout this work, and it would be far lesser without them.

Heartfelt thanks must also go to Matt Smith, my former supervisor, whose comments on an early version of this manuscript proved invaluable. Similarly, I am grateful to the two anonymous reviewers who provided feedback on the first draft. While I have not been able to implement every one of their many suggestions, their advice gave me much to think about, and it encouraged me to make several important revisions. I must apologise to the second, however, for being unwilling to moderate my political stance!

The majority of the writing here was completed during the COVID-19 pandemic, and for the duration I have been in a position of relative economic precarity. With that in mind I would like to acknowledge the assistance of The Society of Authors, which awarded me an Author's Foundation grant in 2021. This financial support was much needed, and it allowed me to dedicate several months to full-time work on the manuscript. I would highly recommend that anyone in a similar situation look into the funding streams this organization runs for struggling writers.

For the duration, I have also worked without access to the resources of a university library. This sometimes made it very challenging to find the books and journal articles that I required. I am therefore grateful to Aileen Lichtenstein, Cynthia Tang, Lauren Young, and Viktor Jörgens, each of whom were able to send me material that I may otherwise not have been able to access. Thanks are also due to Alexandra Elbakyan for all of her assistance.

The staff at Polity made the process of bringing this book to fruition as painless as it could be, and I am grateful to each and every one of them. Thanks in particular should go to my editor Julia Davies for her consistent support, enthusiasm, and patience over the last few years. Lindsey Wimpenny also deserves particular credit for her work researching the images that appear in the following pages.

Writing can be a lonely occupation, and I am forever grateful to the friends who have kept me going through both the highs and the lows. Thanks, then, to Roisin Convery, Patrick Foley, James Turner, Jed Howlett, Catherine Smalley, Michael Gordon, Nicola Cacciatore, Erin Stewart, and those I have almost certainly forgotten to name. I do not get the chance to see all of you as often as I might like, but every one of you has helped me, in one way or another, finish this project – even if only by allowing me to forget about it for a time!

As ever, thanks to my parents, Stephen and Barbara, who have always had my back. I cannot begin to express my gratitude for everything that they have done, and everything that they continue to do.

Most of all, thanks to Alyssa McGrow – without a doubt the best person I have ever known. Her consistent enthusiasm for this project pushed me to keep going on even the most difficult days, and her willingness to constructively challenge my ideas from new perspectives has helped shape the following pages more than she can possibly know. Without her love and support, there is a very good chance that this book might never have existed at all.

Preface

For the first two decades of my life, I gave little thought to diabetes, and much less to insulin – the medication used to treat it. What little I did know about the subject came from the media and occasional comments from friends and family members. I had already left school before I met someone who had been touched by the condition – or, more likely, someone who openly discussed it. As a teenager, I spent some time working in a small charity bookshop in Newcastle upon Tyne. One day, one of my older colleagues arrived for his shift ashen faced. He had, he said, been diagnosed with diabetes, and the news had left him – a man in his forties – visibly shaken.

At the time, I did not know what to say. Watching him struggle to adapt to his new life, wincing as he performed insulin injections in the break-room, seemed only to confirm what I had learned from sensationalist headlines and bleak fictional portrayals. Diabetes seemed terrifying. It scared me, and I could not imagine how I would react were it to happen to me.

As it turned out, I was closer than I would have liked to finding out. As a child, I was asthmatic, and, even though the condition had become much milder by my late teens, I still attended routine review appointments for it. In 2009, my GP surgery mistakenly sent out the wrong form with my clinic invitation letter – it asked that I bring in a urine sample. I was puzzled, but I did as it asked.

When I arrived, the nurse was equally confused, but, seeing as I had the sample with me, she decided there was no reason not to check it out. When she tested it for sugar, it came back as a strong positive. Suddenly concerned, she insisted on taking some blood. When the results came back from the laboratory a few days later, they showed that my blood sugar level was significantly elevated. After some confirmatory tests over the next few months, the diagnosis was clear. I had diabetes too. Lucky me.

At only 19 years old, this came as a shock. I quickly learned why my colleague at the bookshop only a few years before had found the experience so difficult. Insulin therapy involved a great deal more than just doing injections – though they too took some time to acclimatize to. Much more challenging was what it did to my sense of self.

Before my diagnosis, I had enjoyed going on long hikes through the wilderness. Sometimes, I would camp in remote spots, miles from the nearest town. Doing so gave me a feeling of independence that was lacking in the rest of my life. In a naïve way typical of adolescence, I liked to imagine that I could, should the situation demand it, be self-sufficient. The idea that I could ever have lived off the land was, of course, pure fantasy. Nonetheless, the idea that I *might* was, at that time, an important part of my self-image and identity.

Diabetes, however, seemed to make this laughable. I now had to carry insulin and other equipment, and my doctors had explained how important it was to keep my blood sugar levels as stable as possible. If I did not, I might develop serious complications down the line, and in the short term might pass out or even go into a potentially life-threatening coma. This all seemed challenging enough to deal with at home, let alone on some rainy hillside in the Lake District. More importantly, I was now completely reliant upon a pharmaceutical product – and the healthcare infrastructure that provided it. Even a few days without access to insulin could, I was told, prove fatal. Any notion that I might somehow be capable of surviving in the wild for weeks, months, or longer now seemed absurd.

In short, diabetes threatened the very core of my being by stripping away part of my identity and replacing it with something that was, somehow, not me. If the things that made me recognize myself were gone, then who was I? That, especially for someone so young, was not an easy question to answer.

When, years later, I read Australian novelist Peter Corris' memoir of his life using insulin, I immediately recognized the same anxiety.[1] Imagining people with diabetes as weak and impotent, Corris rebelled against his diagnosis for much of his early life. The subjective meaning he attributed to his condition made him reject the label as he struggled to reassert his sense of identity in light of his diagnosis.

This process of reconstitution in light of illness has been described as 'biographical work' by scholars.[2] When someone is diagnosed with a chronic disease, their 'original self' is synthesized with the challenges and expectations, both biological and otherwise, implied by their condition. People need not become their diagnosis – as some worry they might – but it will inevitably transform them into something new, one way or another. Corris, for example, took up smoking and drank heavily in an effort to push his condition away, but that rebelliousness was in itself a product of it.

This interpretation made me re-evaluate my experience. Diabetes forced me to engage in 'biographical work', certainly, but I came to understand that my early belief – that something had been taken – was flawed. There is little that I *absolutely* cannot do, only more risk factors to acknowledge. Whenever I choose to do something, I must consider how it might affect my blood sugar, and decide whether or not to proceed based on my own subjective priority. A diagnosis of diabetes does not threaten the self. Rather, it reveals it.

Insulin therapy is (currently) a lifelong treatment. It insists that those undergoing it refine their personal value systems on a daily basis. They cannot stay passive because it is they who must do the business of treatment every day, incorporating it into their lives as a whole. Sometimes they may choose to do so as healthcare professionals recommend, other times in ways that they would not approve of. In either case, insulin illustrates their rich subjective world. If anything, it reifies the humanity of those with diabetes in all of its complexity and messiness.

This is also, however, a distinctly political substance. Today, it has never been more so. Many of those who need it, like me, need it absolutely. Without it they simply die and too often they are allowed to. Insulin can tell us a great deal about the individual, but it reveals as much about the collective subjectivities of our societies and the ideological frameworks that sustain them.

Insulin saved my life. Over the last decade and a half, I have become rather well acquainted with it. This subject is, as a result, very personal to me, as I am sure it is to countless others. That will be reflected in the following pages, and I make no claims to neutrality – quite the opposite, in fact. In any case, its hundred-year story is fascinating in its own right, and should be of interest to all of us, familiar with diabetes or not.

Introduction: What Is Insulin and Why Does It Matter?

In 1892, the *British Medical Journal* (*BMJ*) published a sobering account of diabetes mellitus in a boy of ten. The child, referred to only by his initials 'J.A.C.', had been admitted to St Thomas' Hospital in London after 'complaining of dryness of the throat and mouth, of being very thirsty, troubled with cough, and of passing more water than natural ... [and] also, though eating heartily, growing thin'. When he was examined, the 'flushed, anxious-looking' boy's urine 'gave immediate evidence of abundance of sugar'. House physician Wilford Watkins-Pitchford considered the situation grave. He immediately placed his patient on a limited diet of milk, water, eggs, green vegetables, and biscuits, and ordered that he be given morphine every three hours. The treatment, however, did little. 'J.A.C.' deteriorated rapidly. Over the following twenty-four hours, he produced a full ten pints of urine, 'passing it with great regularity every hour of the day and night'. Three days later, his condition had worsened significantly. His emaciation had become extreme, and he had begun to hyperventilate and vomit. By the evening, the doctors considered him 'evidently moribund'. He no longer asked for water and, while still apparently conscious, 'had not strength enough to speak'. He died the following day.[1]

Cases like that of 'J.A.C.' were particularly traumatic before the twentieth century. Not only did they often involve otherwise healthy children and adolescents, but those affected usually died a miserable death. To make matters worse, no one could explain what was happening to them with any conviction.

'J.A.C.' was almost certainly experiencing what we would now call Diabetic Ketoacidosis (DKA). Today, he would almost certainly have been given insulin immediately, and there is a good chance that he might have survived to tell the tale. In 1892, however, this was not an option. He did not stand a chance of recovery.

Insulin and Diabetes

So what exactly is insulin – this incredible substance that might have saved 'J.A.C.'s life? Today, it is thought of primarily as a pharmaceutical product, but it is not a drug in the same sense as, say, an antibiotic. Instead, it is a hormone – something that we all need to live, and which most of us make for ourselves. Cells found in the islets of Langerhans, within the pancreas, pump it out as necessary throughout our lives, and it plays an important role in ensuring that we are able to use the energy that we get from food.[2]

When we eat carbohydrates, they are broken down into sugars. When this enters the bloodstream, insulin is the substance responsible for making sure that it gets into our cells, which then burn it as fuel. In simple terms, it acts as a key – prompting our bodies to consume what they need to function.

However, in some, like 'J.A.C.', the insulin-producing islet cells stop working. This happens when the body's immune system mistakenly identifies them as a threat and attacks, impairing their function and eventually stopping it altogether. It is still unclear what exactly triggers this, and, while such cases were historically associated with younger and thinner people, it can happen to anyone.

With insufficient insulin, the body's cells are unable to absorb enough energy and, with nowhere to go, sugar begins to build up in the blood. As less insulin is produced, the energy deficit grows. After a point, this produces physical symptoms. People affected feel the need go to the toilet more often as excess sugar is removed via the urine. This leads to dehydration, and a persistent thirst that cannot be quenched.

Symptoms continue to intensify until the amount of insulin being made is no longer enough to meet even basic requirements. At this point, the body's cells begin to effectively starve, and exhaustion sets in. Protein and fat tissues are broken down to function as an emergency energy source, and visible wasting begins to occur – sometimes over an alarmingly short period of time.

Breaking down these alternative tissues floods the blood with ketones, increasing its acidity and producing a characteristic smell of acetone on the breath. Before long, this leads to DKA, which can be extremely dangerous. It can cause vomiting, stomach pain, breathing difficulties, and disorientation, and, left unchecked, often ends in coma and death.[3]

Without insulin treatment, this process is universally fatal. The exact prognosis varies, as the islet cells are destroyed only gradually, but absolute insulin deficiency leads to death within a matter of days. By the time serious symptoms begin to appear, a considerable amount of damage has already been done.[4] Before insulin, it was rare for anyone to survive beyond a few years following diagnosis.

Today, we would describe this condition as type 1 diabetes mellitus (T1DM). However, this terminology is not particularly old. Before the twentieth century, and for some time into it, diabetes was often considered a singular condition, and cases were distinguished only by symptomatic intensity. If they made a distinction at all, physicians generally referred to T1DM as 'severe', 'juvenile', or, later, 'insulin-dependent' diabetes.

The language we now use was adopted widely only after 1979, following the publication of an influential paper by the USA's National Diabetes Data Group (NDDG). This suggested that diabetes be understood not as one condition, but as a 'genetically and clinically heterogeneous group of disorders that share glucose intolerance in common.'[5]

While 'severe' cases of diabetes seemed uniformly deadly, it had always been obvious that some people seemed to be affected more violently than others. Even as 'J.A.C.' died, others – most of them older and/or larger – avoided such a rapid downwards trajectory. Some even survived for years or longer with relatively manageable, intermittent symptoms.

In 1840, for example, English doctor C.R. Bree described the case of one 53-year-old man who had come to him in July 1835 complaining of 'great weakness'. He was not emaciated, but he was passing twelve to fifteen pints of sweet urine daily, had a great appetite, and could not quench his thirst. Bree recommended a few dietary adjustments, and by the beginning of August the man felt much better. His symptoms had not disappeared completely, but they had become tolerable, and he lived for another three years before dying of dropsy.[6]

Bree's account almost certainly describes what we would now call type 2 diabetes mellitus (T2DM). Like T1DM, this is a complex condition with no single cause. Age and obesity are considered well-established risk factors for developing it, but they are far from the whole story, and it too can affect anyone. Olympic gold medal-winning rower Steve Redgrave, for example, was diagnosed with T2DM at only 35, despite being a world-class athlete in active training.

People who develop T2DM continue to make insulin, though their islet cells do sometimes show evidence of deterioration. The destructive autoimmune response present in T1DM is absent. Instead, this condition is characterized by insulin resistance. In those affected, the islet cells continue to function, but the insulin they make is far less effective at doing its job. Enough sugar is usually processed to meet basic needs, but much is wasted, and it begins to accumulate in the blood. Minor insulin resistance is often completely asymptomatic, but once blood sugar becomes sufficiently elevated it does become noticeable. As the body flushes the excess sugar through the urine, thirst and dehydration develop and, as a significant amount of energy finds its way into a toilet bowl before it can be used, this is often accompanied by lethargy, and sometimes weight loss.

Because insulin, albeit not enough of it, is present in people with T2DM, they are not usually as susceptible to DKA – though it can occur.[7] However, their condition is not benign. In addition to the uncomfortable, frustrating, and sometimes socially embarrassing symptoms, persistently high blood sugar levels can be devastating to the fabric of the body. People with untreated T2DM are significantly more likely than most to develop both heart and kidney disease, suffer strokes, and lose their sight. They also sometimes experience painful nerve damage and sexual problems such as erectile dysfunction, while miscarriages during pregnancy are more common.

Lifestyle changes can in some cases act to restore sensitivity to the body's own supply of insulin in cases of T2DM, and pills like Metformin can help further, but this is not always sufficient to keep blood sugar beneath the threshold at which damage might occur. While it is not so immediately essential to life itself, injected insulin can play a vitally important role in maintaining health in such cases.

Insulin has been available as a medication since 1922, but it has not trivialized diabetes. As paediatrician and historian Chris Feudtner puts it, its introduction to medical practice was akin to 'a Greek myth of rebirth turned ironic and macabre'.[8] It does not permanently *cure* anything. It only replaces, or, in the case of T2DM, supplements, something usually made by the body. As a result, those who need it must take it for life.

Complicating things further, insulin is not 'fire-and-forget' medication like, for example, an aspirin. If it is not taken in the right amount and at appropriate times to balance out demand, blood sugar levels can rise or fall to dangerous levels. Too much insulin in the body can lead to hypoglycaemia (low blood sugar) which, in extreme cases, can cause unconsciousness and even death. Too little, and sugar levels become dangerously elevated, damaging the organs, nerves, and blood vessels and, left unchecked, perhaps even leading to DKA.

Insulin has, however, transformed the experience of diabetes. Those who are able to access it are given a reprieve that their predecessors were not. Long-term therapy, however, brings its own challenges – some of them very demanding indeed. But we are getting a little ahead of ourselves. It is impossible to discuss the history of insulin as a pharmaceutical tool without first considering the circumstances that led to its 'discovery'.

A (Very) Short History of Diabetes Before Insulin

While it has been given a variety of names throughout history, diabetes is likely ancient. The term itself was first used by the classical Greeks, and is derived from their word for 'siphon', or 'to pass through', reflecting the huge amount of urine produced by those affected. 'Mellitus' – Latin for 'sweet like honey' – was later appended to indicate the characteristic abundance of sugar that it contained.[9]

Serious illnesses with symptoms paralleling those experienced by
'J.A.C.' in late nineteenth-century Britain have been described across
cultural boundaries throughout recorded history. The Ebers Papyrus,
an Egyptian document dated to around 1552 BC, contains reference,
albeit quite vague, to 'urine which is too plentiful'.[10] Ayurvedic texts
from the Indian subcontinent are more precise, identifying *madhumeha*
– 'honey-urine' – as a dangerous condition characterized by the passage
of great quantities of sweet urine, unquenchable thirst, and bad breath.[11]
In imperial China, *xiāo kě* (消渴) – 'wasting-thirst' – was associated with
frequent urination, extreme thirst and hunger, and rapid weight loss
well over two thousand years ago.[12]

The most complete, and likely most cited, classical account comes
from the second century Greek physician Aretaeus of Cappadocia, who
practiced in what is now eastern Turkey. Describing the symptoms and
progress of those affected in excruciating detail, he wrote that:

> The patients never stop making water, but the flow is incessant, as
> if from the opening of aqueducts . . . the patient is short-lived, if the
> constitution of the disease be completely established; for the melting is
> rapid, the death speedy. Moreover, life is disgusting and painful; thirst,
> unquenchable.[13]

These old accounts are often incomplete and fragmentary, but the key
elements remain strikingly consistent across vastly disparate societies
and time periods. Most early accounts of diabetes appear to describe
T1DM. This is not surprising. Such cases were rare enough to be
notable – Aretaeus thought the condition 'not very frequent among
men' – while the intensity of symptoms made them difficult to ignore.

In 1679, however, Thomas Willis, court physician to Charles I,
argued in his posthumously published *Pharmaceutie Rationalis* that diabetes had become considerably more prevalent than it had once been.
Apparently something of a trendsetter, Willis blamed this squarely on
overindulgence:

> The Diabetes was a Disease so rare among the Ancients, that many
> famous Physicians have not so much as mentioned it, and Galen never
> knew above two that were troubled with it; but in our Age, that is given so

much to drinking and especially to guzzling of strong Wine, we meet with very frequent, not to say daily examples and instances of this Distemper.[14]

Willis believed that diabetes could effectively be controlled, writing that 'this disease at first beginning is oftentimes easily cured; but when grown strong in a man, very seldom and with great difficulty'. He gives the example of one 'noble Earl . . . in the very vigour of his Age', whose symptoms were successfully eased following treatment:

[He] became much inclined to excessive urination; and when for several Months he had been used ever now and then to make great quantities of water he at last (it seemed) fell into a diabetes that was obstinate and strong and almost desperate. For besides that in the space of 24 hours he voided almost a Gallon and a half of limpid, clear and wonderfully sweet water, that tasted as if it had been mixed with Honey; he was likewise troubled with an extraordinary thirst, and as it were, an Hecktick Fever, with a mighty languishing of his spirits, weakness in his limbs, and consumption of his whole body.[15]

After being given several plant-based remedies and placed on a restricted diet, the Earl gradually recovered. Within a month, he was 'almost quite well' – his urine no longer tasted of sugar and it 'did not much exceed the quantity of that liquid matter which he took in'. Eventually, after 'recovering his usual tenure of spirits and strength he returned to his former diet.' Tellingly, his symptoms returned intermittently for the remainder of his life:

But yet the disposition to this Distemper did not so totally leave him but that afterwards, oftentimes, through disorders in his Diet and perhaps irregularities in the seasons of the Year, being inclined to a relapse he made water at first in great quantities and then clear and sweet with thirstiness feverishness and languishment of his spirits [sic].[16]

This certainly does not seem to reflect the terminal condition described by Aretaeus and other early writers. Willis was an extraordinarily successful physician whose wealth and status meant that most of his patients came from the elite ranks of Stuart society. As he explicitly pointed out, this demographic was not known for its temperance, and obesity was not uncommon. Moving in these circles, it is very likely that the majority of Willis' cases involved T2DM.

Willis' powerful patients in the mid-1600s were a small minority unrepresentative of the general population of Britain. However, by the nineteenth century most physicians did tend to agree that diabetes had become commonplace. Many echoed his claim that it was a disease of affluence, and most of those seeking treatment in this period certainly did come from the more prosperous sections of society.

At the same time, others were beginning to highlight that it seemed to have become significantly more survivable than most historical accounts suggested it should be. In 1866, Scottish physician George Harley wrote that:

> At one time, when a patient was said to have diabetes, it was considered tantamount to saying his days were numbered. As our knowledge has increased, however, we have learned that although it is but seldom possible for us to eradicate the disease, we can, nevertheless, so mitigate its effects as not only to prolong the life of the individual, but to render it a tolerable, if not even an agreeable one.[17]

Harley – like many of his contemporaries – was eager to credit the effort of the medical profession with any apparent improvement in prognosis, but it is quite likely that he was actually just observing the consequences of broad demographic changes within Victorian society. Thanks to the Industrial Revolution and the global hegemony of Britain's vast colonial empire, by 1866 there were considerably more people in the UK that would today be considered at high risk of T2DM than there had been a hundred years before.

The medical profession did have some success at ameliorating the symptoms of these new patients. In 1797, John Rollo, then surgeon-general of the Royal Artillery, outlined a low-carbohydrate, high-fat 'animal food' diet that appeared to have some positive effect. One 57-year-old general officer under his care, for example, had been diagnosed with diabetes approximately three years prior, and regularly produced 'ten or twelve pints [of sweet urine] in . . . twenty-four hours'. When consulted, Rollo advised him to adopt a heavily restricted diet that sounds singularly unappetizing: blood pudding for lunch, 'game and old meats which have been long kept, and, as far as the stomach may bear, fat and rancid old meats' for dinner, and bread and butter for both breakfast and supper, washed down with a mixture of milk and lime water.

Impressively, the man soon did seem to improve. After some time on Rollo's diet, he was no longer so thirsty and he was passing between two and four pints of urine daily. He even appeared to be regaining his strength. Following his apparent recovery, the officer in question – who had 'very great impatience under restriction' – returned home to Portsmouth. After receiving permission from another physician to 'eat what he pleased, and to drink wine' following a bowel complaint, his diabetic symptoms soon returned. Rollo, with some justification, took this as vindication for his approach to treatment.[18]

Rollo's restricted diet was highly influential, and while other physicians tinkered somewhat with the formula, most maintained the principle that diabetes was best treated by proportionally reducing carbohydrate in favour of greater quantities of meat and fat. This method of treatment remained common into the twentieth century, probably because it did, on at least some occasions, seem to work – particularly for those we would now recognize as having T2DM.

Nonetheless, no treatment could do anything for 'severe' cases, who continued to die quickly. By the middle of the nineteenth century, the stark differences in outcomes between people diagnosed with diabetes prompted some speculation that at least two distinct conditions might be responsible. In 1863, Manchester physician Daniel Noble, for example, asked whether it might be 'possible to establish pathological distinctions in cases of diabetes, according to their *origin*, the *course of the symptoms*, and their *curability*'.[19]

In 1866, George Harley had (correctly) surmised that 'saccharine urine is not itself the disease, but merely the most prominent sign of ... a variety of morbid actions'. He distinguished between diabetes he thought was caused by 'excessive formation' on one hand, and by 'defective assimilation' on the other:

> In those resulting from *excessive formation* the patient is not necessarily emaciated and weak. He may, on the contrary, look both fat and ruddy

– appearing, in fact, to be the very bloom of health . . . In this class of patients it is not until the disease has made considerable inroads on the constitution that there is any marked emaciation.

In the second class of cases, on the other hand, namely, those resulting from *defective assimilation*, emaciation is one of the earliest and most prominent symptoms, the loss of the flesh being often very marked before the nature of the disease is detected.

Harley's descriptions quite clearly reflect T2DM and T1DM respectively, but his theory was entirely speculative.[20] He was only one voice amongst many, and, throughout the nineteenth century, doctors came to very little agreement as they sought to explain the root cause of diabetic symptoms and draw lines between different 'types' of the condition – if they believed any such division could be made at all.

One major barrier to any real consensus about what exactly diabetes was, and why some people seemed to become far sicker than others, was the ambiguity of its origin. Until the late nineteenth century, physicians argued at length about where to locate its 'seat' in the body. By the late nineteenth century the pancreas had – correctly – been implicated several times, for example by French physician Étienne Lancereaux, but an undeniable link remained elusive.[21]

This changed in 1889, when German physicians Oskar Minkowski and Josef von Mering removed the pancreas of one of their experimental dogs for study. Following this procedure, they later reported, the animal had begun to relieve itself uncontrollably all over the laboratory floor and had died soon after. On examination, its urine was found to be full of sugar. It appeared to have spontaneously developed the kind of quickly fatal symptoms seen only in the most 'severe' cases of diabetes, though precisely why neither could say for sure.[22]

The concept of an insulin-like substance, in the very broad sense, can probably be traced to French-Mauritian physiologist Charles-Édouard Brown-Séquard. From the mid-nineteenth century, Brown-Séquard argued that various glands throughout the body produced 'internal secretions' that each played an essential role in maintaining health.[23] Building on this, he later went on to suggest that, if extracted, they might have therapeutic value.[24] His ideas were controversial at first, and his belief that injections of bull semen might rejuvenate old men who had long since lost their strength and virility was predictably mocked by his peers, but in principle he was correct – he had, in fact, identified the group of substances University College London's Ernest Starling would, in 1905, give the collective name 'hormones'.[25]

By the mid-1890s Brown-Séquard's theory had become widely influential.[26] One reason for this was that the work of other researchers seemed to lend it clear support. Minkowski and von Mering's experience, for example, had after all shown that, without a pancreas, dogs would become sick and die in short order. In 1891 the idea gained even further traction when George Redmayne Murray reported that he had been able to successfully treat a woman with myxoedema – advanced hypothyroidism – by injecting her with extracts from the thyroid glands of a sheep.[27]

Murray's success prompted a great deal of optimism amongst the medical community, particularly when it became apparent that the extract could be given orally without issue.[28] By 1893 Brown-Séquard clearly considered his ideas thoroughly vindicated, highlighting diabetes as one example of a condition he thought would – like myxoedema – benefit from treatment with organ extracts. 'There cannot be a doubt now', he claimed, that the pancreas, 'like the testicles, the ovaries, [and] the kidneys, has an internal secretion, which is even more important than its external one [i.e., digestive fluids]'.[29]

He was, of course, correct, and was describing the hormone Belgian Jean de Meyer would christen 'insuline' in 1909.[30] However, initial attempts to use pancreatic extract were not promising.[31] While he defended his 'internal secretion' theory, Brown-Séquard was forced to concede that 'while injections of the pancreatic liquid have been useful . . . no case of cure, to my knowledge, has been recorded'.[32]

By the end of the nineteenth century, growing attention to the 'internal secretion' of the pancreas discouraged many physicians from believing that multiple varieties of diabetes existed. In 1896, for example, Canadian physician William Osler suggested that while both 'acute' and 'chronic' forms existed, there was 'no essential difference between them, except that in the former the patients are younger, the course more rapid, and the emaciation more marked'.[33]

By 1919, New York's Frederick Allen – widely acknowledged as one of the foremost global specialists on diabetes – felt comfortable arguing that it was one thing and one thing alone: a condition resulting from 'a deficiency of the internal secretion of the islands of Langerhans'.[34] He had little time for those who continued to suggest otherwise, brusquely dismissing the work of Lancereaux, for example, as 'generally discredited'.[35] The status of the islet cells in the pancreas 'as the seat of the specific diabetic disturbance', he declared, 'is now as firmly established as any fact in physiology or pathology'.[36]

Allen believed that high blood sugar was more than an indicator that the pancreas was producing insufficient 'internal secretion'. He also thought that it could make things worse by overworking the islet cells, and in doing so causing them to degenerate further.[37] The greater the burden placed on the organ, he argued, the more rapid this process would be. A 'mild' case would, left unchecked, always become a 'severe' one, and a 'severe' one would eventually become terminal.

While he understood diabetes as a singular condition, Allen did recognize that, cruelly, children seemed to develop life-threatening symptoms more rapidly than their older counterparts, insisting that 'every case . . . in a child calls for the most careful treatment from the earliest possible moment'.[38] As his long-time ally Elliott Joslin wrote in 1916, however, the 'melancholy fact' was that 'where diabetes appears in its most severe type, as in children, coma is its expression'.[39]

By the 1920s, Allen had become well-known for his radical approach to dietary treatment. If deterioration was caused by undue pressure on the islet cells, he thought, then normalizing blood sugar through nutritional restriction would reduce that stress and stall the progression of diabetes. Dietary regulation was by this point nothing new, but Allen deviated from his predecessors by advocating a particularly aggressive strategy, and claimed that remaining consistently underweight was beneficial. His method involved, to put it mildly, radical under-nutrition. Where those undergoing treatment in the nineteenth century might have complained about being asked to cut bread, wine, and sweet foods from their diets, their prescriptions would have looked positively indulgent to Allen, who not only demanded drastically reduced carbohydrate intake, but substantially less food altogether.

It is not at all surprising that Allen's strategy was pejoratively referred to as the 'starvation diet', or that the name stuck. While his exact prescription was tailored to each individual, patients at New York's

Rockefeller Institute, where Allen worked from 1913, were often subject to extreme nutritional restriction alongside periods of almost total deprivation, particularly if they were young and/or admitted with serious symptoms.

One 13-year-old boy, for example, was hospitalized with shortness of breath and disorientation on 12 February 1917, and was quickly found to have extremely sugary urine. On his first day of admission, he was given only 20 ml of soup, which was increased to 60 ml – plus 40 ml of coffee – for the next four days. Following this period of fasting he was moved onto a 'standard' diet, which still amounted to only 240 calories daily for around two weeks.[40]

Unsurprisingly, Allen was not popular amongst his patients. In addition to his harsh approach to treatment, he was every inch the stereotype of the paternalistic early twentieth century physician – a severe, humourless man who valued hard work, temperance, and discipline above all else. Combined with an absolute faith in his own style of diabetes management, he held a palpable contempt for those who breached his 'rules'. As far as Allen was concerned, insubordination was one of the primary causes of death in those under his care: he separated fatalities into two groups, those attributed to 'disobedience of patient', and those in which 'failure of treatment' was responsible.[41] He viewed the moral character of those under his care as an integral part of management, and felt that those who 'failed' in their duty brought disaster upon themselves:

> The patients who died from breaking diet were not driven to desperation by hunger or suffering. They were generally not the ones who had to endure the greatest privations. They were rather the ignorant, the careless, the weak-willed, the neuropathic, and others who would not have been faithful to any restrictions no matter how mild.[42]

It is difficult to imagine how oppressive this must have made life on the wards, but Allen's case records give some indication as to the kind of environment his patients endured while accommodated at the Rockefeller, particularly if he decided that they were, for whatever reason, lazy or untrustworthy. In one case, when he could not explain a 'barely perceptible' resurgence of sugar in the urine of one visually impaired and bedridden 12-year-old, he interrogated the boy, who eventually admitted to breaking his diet:

> It had seemed that a blind boy isolated in a hospital room and so weak that he could scarcely leave his bed would not be able to obtain food surreptitiously when only trustworthy persons were admitted. It turned out that his supposed helplessness was the very thing that gave him opportunities which other persons lacked . . . the attempt to evade the strict watch kept over him appealed to him as a sort of game or battle of wits, so that he even took things for which he had no real desire.

The boy, who had been given fewer than seven hundred calories daily for almost a week, and on some days considerably less, had been eating 'tooth-paste and bird-seed, the latter being obtained from the cage of a canary which he had asked for'. Allen did not consider that this behaviour might reflect the desperation of an unbearably hungry child unwilling to openly challenge his doctor's instruction, and appears instead to have concluded that the whole episode was some kind of infantile game.[43]

Despite all of this, he did not lack for prospective clients. This speaks to quite how hopeless 'severe' cases of diabetes must have seemed. Why else would anyone have subjected themselves or their loved ones to this kind of ordeal? The response amongst the medical community was also enthusiastic. By the First World War Allen had become highly influential on both sides of the Atlantic. In 1917, British physician Otto Leyton eagerly endorsed his method, which he described as 'the modern treatment of diabetes mellitus'.[44] George Graham used a similar approach at London's St Bartholomew's Hospital in 1921, making it even more onerous by insisting his patients remain in bed for most of the day. He justified this by suggesting that it might make dietary restriction more bearable, but also admitted that it was first implemented to ensure that his staff could keep close tabs on them.[45]

It is difficult to accurately assess how effective this harsh treatment was. We do know that of 44 patients under Allen's care at the Rockefeller in 1915 – not all of whom will have had T1DM – almost half were dead by the end of 1917.[46] Some had died of complications related to diabetes. Others simply starved. Even Allen admitted that it could only delay, not prevent, the inevitable.[47] When, years later, one physician described early twentieth century treatment as little more than 'a counsel of desperation', the assessment seemed little exaggeration.[48]

One of Allen's most famous patients was a young girl called Elizabeth Hughes. Hughes had been diagnosed with diabetes in 1918 at the age of eleven, and she came under his care a few months later in early 1919. She was placed on a diet typical of Allen's practice – 889 calories per day. While she personally did not, with some justification, much enjoy the doctor's company, Allen considered her a model patient. As her health waned, she continued to abide strictly by her prescribed treatment. Nonetheless, few expected her to live into adulthood. Even Allen must have been pessimistic, and his stated mission to wring as much life out of his patients as possible in the hope that they might live long enough to see a cure must have seemed cold comfort:

> Final attention must again be called to the limitations inherent in every dietetic treatment. It affords only rest of a weakened function, when a stimulus is often needed. Essential progress must take the direction of supplementing the negative and passive therapy with a positive and active force. The knowledge of diabetes is advancing rapidly enough that even the patient whose outlook seems darkest should take courage to remain alive in the hope of treatment that can be called curative.[49]

In the early months of 1922, however, it seemed that just such a 'positive and active force' might finally be within reach. Promising news had begun to arrive from Canada. A team of researchers in Toronto, led by Frederick Banting, claimed to have successfully produced an effective pancreatic extract that could stop diabetes in its tracks. The substance – insulin – seemed almost miraculous: it appeared to have brought those children it was tested on back from the very brink of death (see Figure 1).

As the daughter of veteran American politician Charles Evans Hughes – a former Governor of New York who had only narrowly lost to Woodrow Wilson as the Republican nominee in the 1916 presidential election – Elizabeth was as well cared-for as any child of her era could have hoped to have been. She was able to rest in comfortable surroundings and was attended to daily by a private nurse.

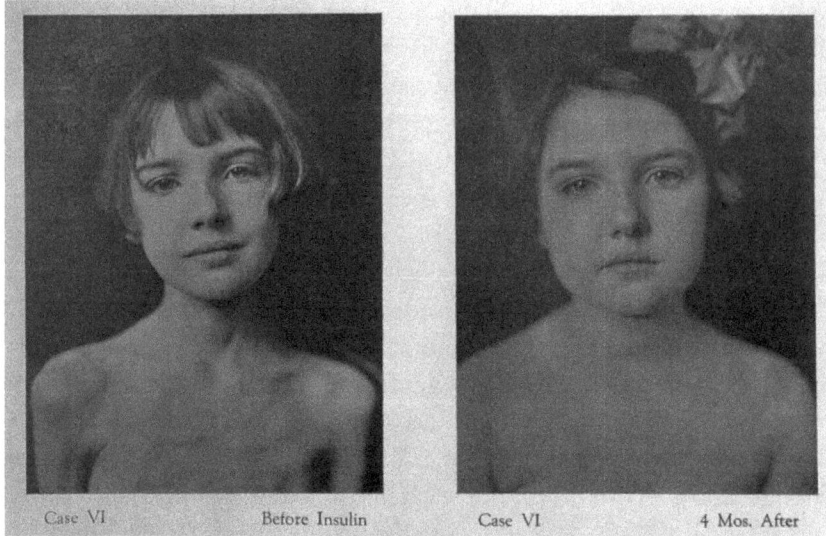

Figure 1. *A young girl before and four months after insulin treatment,* c.1922
Source: Wellcome Collection

By the summer of 1922, however, she was in increasingly poor health.
Horrendously emaciated, she weighed only 40 lbs (18 kg) and was so
exhausted that she struggled to walk. However, she was fortunate to
be born into such a high-status family. After repeated requests by her
parents, Banting agreed to take her on as a private patient.

Hughes travelled to Toronto and received her first injections of insulin in August, and afterwards recovered quickly. The sugar in her urine
disappeared, and within a fortnight Banting was allowing her a diet of
around 2500 calories per day, leaving the starved girl overwhelmed
with relief. At the end of September, she wrote elatedly to her parents:

> As you know I am simply bursting to see you and can hardly wait for
> you to actually see with your own eyes what I'm eating nowadays, for
> if you didn't I declare you'd think it was a fairy tale. I know you will
> hardly know me as your weak, thin daughter, for I look entirely differ-
> ent everybody says, and I can even see it myself.[50]

Hughes soon left the hospital, and ended up living into her seven-
ties. Within a matter of weeks, she appeared to have recovered almost

completely with little to show for her ordeal. Insulin seemed an almost unbelievable achievement, and a hundred years later this still seems a fair assessment. To date, it has saved countless lives, and, should we let it, it will save countless more.

Insulin at One Hundred

Hughes is a good place to finish the introduction of a book celebrating a century of insulin. She was one of its most high-profile early recipients and her story has become well known, but she also highlights its darker implications. Every human body requires insulin, but most make their own. For those who no longer do, injections are as vital as food, water, or oxygen. If treatment is stopped, symptoms return quickly and death looms not far beyond. While it is often described as a chronic condition, T1DM in particular is better understood as a persistently treated acute one – each injection only delays deterioration for a few more hours. There is no 'cure'. Insulin only maintains life for as long as the people who need it can get it.

In the 1920s, those who required insulin usually had to buy it. In these early years, when supplies remained stretched, this could be costly. In Canada in 1923, for example, it could cost up to a dollar per day in a country where the average annual income was only around C\$500.[51] Hughes' life was saved as much by her class context as it was by any pharmaceutical product, and while it is difficult not to be moved by her tale of triumph over adversity, it is important to remember that it occurred alongside countless tragedies.

Many people continued to die needlessly of untreated (or insufficiently treated) diabetes after the introduction of insulin to medical practice. Unsurprisingly, most of Banting's private clients came from the upper echelons of North American society. Hughes was the daughter of a statesman. James Havens – who went on to become an accomplished artist – was the son of a congressman. Teddy Ryder must have felt in comparatively privileged company as the child of a humble engineer!

Had she been born poor to an insignificant family, Hughes would almost certainly not have survived. When we take the time to celebrate the 'discovery' of insulin, it is vitally important that we acknowledge the countless working-class Elizabeths who died and were forgotten even as she received her first life-saving injections. This is particularly relevant today. In 2023, over a full century later, many continue to meet this fate for no other reason than simple economic misfortune. Insulin accessibility remains a major issue across the world despite well-established manufacturing operations that produce enough to meet the needs of all who require it. This is a particular issue in low- and middle-income states where poverty is rife, average wages are low, and healthcare infrastructures are both underdeveloped and underfunded, if they meaningfully exist at all.

However, ideology as well as deprivation remains the source of much suffering. The United States is probably the most egregious example of this. Despite being the richest, most powerful, and most culturally influential country on the planet, its political leadership has consistently shown itself to be remarkably unwilling to enact legislation that would provide full access to medical care for all of its citizens. Most working-age Americans rely upon private insurance policies often tied to employment – either their own or that of a family member.[52] Private insurance is often very expensive, and even the more comprehensive schemes demand significant contributions from the individual in the form of deductibles and 'co-pays'. Worse, those in poorly paid (and usually under-appreciated) jobs often do not receive insurance at all. In these cases, they are responsible for purchasing their own coverage, and, given the oppressive cost, some end up going without. The most vulnerable in American society are asked to pay the most, despite being the least capable of doing so. Many simply do not even attempt to access the care they need when they become sick or get injured in the knowledge that doing so might land them in significant debt, even where they acknowledge that the decision puts their health at serious risk.

It is difficult to convey how expensive health coverage can be in the United States to someone with little experience of it. One darkly humorous 2013 webcomic, for example, points out that the central premise of critically acclaimed TV show Breaking Bad – in which a mild-mannered chemistry teacher turns to producing and selling meth-amphetamine in order to pay for the expensive cancer treatment his insurance will not cover – would simply not make sense in countries with functional universal healthcare systems, highlighting how absurd it is that wrestling with medical bills has become a common trope in media produced in the richest country on Earth.[53]

While it is an oversimplification, this is uncomfortably close to the mark. Uninsured, even an emergency ambulance call can run into the thousands of dollars, and treatment itself can cost many times that – particularly for more serious conditions.[54] Today, insulin can cost upwards of $1000 for a month's supply, and even those with insurance find themselves paying thousands each year to meet the requirements of their policies – it is that, or death.

The history of insulin is therefore not the simple history of uncomplicated medical innovation making a once fatal disease survivable. Its first use in 1922 was only the beginning of a much longer story that forces us to engage seriously with questions about the way healthcare is – or, too often, is not – delivered globally, and the implications of leaving the provision of life-saving medication to profit-seeking companies within the broader context of free-market capitalism.

There is more to insulin's story, however. Beyond issues of economics and access, it invites us to reflect on ideas of power, authority, and subjectivity within medicine. In the early twentieth century, doctors approached their insulin-using patients much as they did any other. As the professional, they often assumed an authoritative stance. They quickly discovered, however, that there was no way to enforce this. Day-to-day treatment occurs far from any doctor's office or hospital ward. The final decision on how much to inject and when lies with the user, not with their doctor. Insulin, as a result, transfers de facto power over healthcare to its recipient, whether or not they welcome it. This has important implications.

Individual decisions about treatment in diabetes rarely have immediate life or death consequences. Instead, they impact blood-sugar levels in ways that may or may not contribute to long-term risk factors – for example for kidney disease or nerve damage – or run the risk of hypoglycaemia. With full responsibility in the hands of the person undergoing treatment, insulin therapy as a result becomes a space in which subjective health beliefs and values are interrogated on a daily basis.

Certain lifestyle choices, for example, might threaten overall blood sugar control without any immediate danger, but may potentially increase the long-term risk of complications, and the individual must consistently make value judgements on this basis – deciding how to proceed based on their own subjective needs and broader life-context.

Insulin drags healthcare out of the clinic. It forces us to engage with medicine beyond its professional borders. Someone using it is able to make active choices about which risks they consider worthwhile, regardless of their doctor's wishes. In doing so, they may put themselves at higher risk of long-term complications, but does this necessarily imply a failure of treatment? It demands that we ask these difficult questions – what precisely is 'health' and what makes treatment 'successful'? How do we define expertise in the context of chronic health conditions? Who should have final authority in care and what are the implications of a patient-body able to unilaterally reject professional instruction?

As of 2023, insulin has been used for over a century. This is reason enough to tell its story, but it is also directly relevant to our present time. Accessibility issues in North America often dominate the headlines, but the challenge is international. Rates of diabetes continue to rise globally, particularly in those middle-income countries that are rapidly industrializing and undergoing a demographic transformation not unlike that experienced by the Western European states of the eighteenth and nineteenth centuries. According to the International Diabetes Federation (IDF), 537 million people currently live with diabetes of one kind or another, while the organization expects this figure to grow to 783 million by 2045.[55]

Insulin has never been more relevant. As the accessibility crisis rages on, increased political engagement has for many become an existential necessity. Activist communities highlight the tragic stories of those who have died for lack of health coverage and directly agitate for guaranteed access as part of a broader coalition demanding universal health coverage. The story of insulin is one that highlights the shortcomings of an inherently destructive economic system, and emphasizes at every turn the need for a more redistributive global politics.

Chapter 1

Toronto, 1921–1923

The story of insulin's 'discovery' has become one of the most enduring in the history of medicine. It has been told many times over – most comprehensively by Michael Bliss in his classic work *The Discovery of Insulin* – but a century later it remains just as gripping.[1] With a rags-to-riches character arc, high drama, and a heroic conclusion, it feels almost as if it has been lifted directly from a Hollywood script. As ever, however, the simple narrative hides a far more interesting, if murky, tale, and one that invites us to ask serious questions about the way we understand the process of research and innovation in medicine.

Frederick Banting, so it goes, made for an unlikely protagonist. Born into a relatively prosperous farming family in the countryside around Alliston, Ontario, he was not a particularly academic child and only barely managed to graduate from high school. In 1910, he began to study arts at the University of Toronto but promptly dropped out after failing some of his first-year examinations. He then changed direction completely, sought re-admittance and started a medical degree in 1912.

Banting was better at medicine than he had been at the arts, but he was never a particularly remarkable student. His final year was compressed into only a few months due to the heavy demand for medics on the battlefields of the First World War, and after his graduation in late 1916 he, like most of his cohort, went off to Europe to fight. He took

well to army life, and even earned a Military Cross for bravery under
fire, though not without being wounded by shrapnel. After some time
spent recuperating in Britain, he returned to civilian life in Canada in
1919.[2]

After the war, Banting struggled to find work. Following a brief stint
as a surgical resident at Toronto's Hospital for Sick Children where,
despite his best efforts, he was unable to secure a permanent position,
he decided to try his hand at private practice. In the summer of 1920,
he opened a modest office in London, Ontario. The fledgling business
was, however, a complete failure. Increasingly indebted and profes-
sionally embarrassed, he took on a part-time job as a demonstrator
at the University of Western Ontario – known colloquially simply as
'Western'.[3]

Years later, in a 1929 lecture, Banting tried to put a positive spin on
this period of his life. He was, he suggested, determined to become a
fellow of the Royal College of Surgeons of Edinburgh, and working
at Western meant that he had easy access to the resources he needed
to prepare for the necessary examinations.[4] This was not an unreasonable goal for a man of his age and experience, but he also desperately
needed the cash – throughout the summer and autumn he was making
only around fifty dollars per month, the equivalent of just over C$600
today.[5] When the costs of running his practice were taken into account,
there was no hiding the fact that he was haemorrhaging money.

Unsurprisingly, Banting was miserable. Decades later he still found
it 'impossible to forget the awfulness, the loneliness and the financial
worries' of this period of his life, where 'painting and studying and
teaching' were, apparently, his only pleasures.[6] By late 1920 he was
exhausted and increasingly disillusioned with private practice. Few –
himself probably least of all – could have predicted that he was on the
verge of a dramatic change of fortune.

Banting's 'Eureka' Moment

Towards the end of October, Banting was asked to deliver a talk on
nutrition to a physiology class – specifically on the topic of carbohydrates and the way the body processes them.[7] As a man much more
interested in surgery, he seemed a strange choice. This was far from his

specialist subject, and there is a good chance that some of his students were more up to date on the literature than he was. Nonetheless, he prepared the lecture as requested, finishing work late in the evening on 31 October. Apparently not a fan of light bedtime reading, he then decided to wind down for the night by browsing a copy of the latest volume of *Surgery, Gynecology and Obstetrics*. In it, he came across an article by Moses Barron which, as chance would have it, was on the very same subject he was due to speak about.[8]

Barron's article described several post-mortem examinations. One of these, performed on a 40-year-old man who had died soon after being admitted unconscious to the Minneapolis City Hospital, seemed unusual. A large gallstone had been found completely obstructing the main pancreatic duct – the passageway through which digestive fluid leaves the organ. Most of the pancreas itself was withered and dead. On closer inspection, however, the doctors present noticed that one group of cells – the islets of Langerhans – seemed to remain functional.

This case seemed to parallel the results of several prior experimental animal studies in which the pancreatic duct had been artificially tied closed ('ligated'). Whereas fully removing the organ – as Oskar Minkowski and Josef von Mering had – tended to quickly produce the symptoms of diabetes in animals, this procedure did not. The animals' pancreases did degenerate, but they did not become noticeably ill.[9] This, Barron argued, suggested that it was the islet cells, rather than the pancreas as a whole, that were responsible for 'secret[ing] a hormone directly into the lymph or blood streams ... which has a controlling power over carbohydrate metabolism', and that these cells were able to function regardless of the status of the rest of the organ.[10]

That the islets of Langerhans might be centrally important in producing such a secretion was not a new idea by 1920. French pathologist Gustave-Édouard Laguesse had speculated that this might be the case as early as 1894. However, it is likely that Barron's article was the first time Banting had been exposed to the theory, or at least the first time he had given it any serious thought. Turning in, he had a restless night and slept poorly. His thoughts meandered. But then an idea coalesced:

[It] occurred to me that by the experimental ligation of the duct and the subsequent degeneration of a portion of the pancreas, that one might

obtain the internal secretion free from the external secretion. I got up
and wrote down the idea and spent most of the night thinking about
it.[11]

In simple terms, Banting's theory was as follows: the pancreas was
likely responsible for producing both powerful digestive enzymes (the
'external secretion') and a substance necessary for utilizing carbohydrate (the 'internal secretion'). If the latter could be captured, it might
have some use in reducing blood sugar. Previous attempts to isolate
it had, however, been largely unsuccessful. This, he reasoned, might
be because when the two came into contact, the 'external secretion'
destroyed the more delicate 'internal' one. However, if it was true that
the latter was made specifically by the islets of Langerhans, and if it was
also true that their health was almost completely independent of that of
the rest of the pancreas, then perhaps there was a simple solution. If the
pancreatic ducts were closed, the parts of the organ that made digestive
enzymes would degenerate, leaving only the islet cells behind. From
there, the much sought-after 'internal secretion' could, in theory, be
harvested in relatively pure form, safe from any damage.

Banting's scribbled note, written in semi-legible cursive, is quite
clearly the work of a tired, stressed man in the early hours of the morning. It reads 'Diabetes. Ligate pancreatic ducts of dog. Keep dogs alive
till acini degenerate leaving Islets. Try to isolate the internal secretion
of these to relieve glycosurea [sic].'[12]

Probably for the first time in a long while, Banting woke up full of
enthusiasm. He discussed his theory with colleagues at the university,
but while they were encouraging, he was told that Western lacked
either the resources or specialist expertise to play host to the kind of
experiments necessary to pursue it further. He was not dissuaded. The
personal stakes were too high – this could be a route to serious professional recognition.

One name that did come up frequently in his discussions with colleagues was John Macleod. Macleod was a Scot who had worked at
universities across Europe and North America and was considered
one of the foremost global authorities on the subject of carbohydrate
metabolism. If anyone could say whether the idea had merit, it would
be him. Macleod, as it turned out, had taken a job at the University of

Toronto only a few years prior in 1918, and, in a stroke of luck, Banting intended to be in the city the very next week to attend a wedding.

In many respects it is quite surprising that Macleod decided to speak with Banting at all. The two men had never previously met, and the senior professor was a well-established and highly respected academic who could easily have decided that meeting this insignificant small-town doctor was not worth his time. Nonetheless, he agreed to hear him out. In the event, the meeting went spectacularly poorly – so poorly, in fact, that Banting later remembered Macleod's attention drifting from him to the papers on his desk, which he began to obviously read. Things had not gotten off to a good start.[13]

While he agreed with the basic premise that digestive enzymes might be responsible for destroying the 'internal secretion', Macleod was not impressed. Banting was never a particularly articulate or confident speaker, and his overall knowledge of the topic was superficial.[14] Effectively a layman, he had turned up at the office of a global authority on the subject with glaring gaps in his knowledge, only to claim that he had, in a single evening, devised One Weird Trick that might solve a problem with which esteemed scientists had wrestled unsuccessfully for decades. Charitably, he seemed naïve; less charitably, arrogant.[15]

Despite these poor first impressions, Banting somehow won Macleod around, or at least left him sufficiently intrigued to take a chance. It is impossible to say what precisely the turning point was, but despite his reservations he clearly felt that that something in the theory was worth pursuing. The professor remained relatively unenthusiastic, but agreed that it was 'worth trying' and indicated that he would support some preliminary investigations.[16]

Macleod, however, warned that this was no small undertaking, and that in all likelihood it would come to nothing. Did Banting really want to give up his life in London for what would probably be an exercise in futility? He returned home and mulled over the issue for several months. It was obvious that pursuing something so risky was a huge gamble. In London, he had a foot in the door at Western, and trade had finally begun to improve, but there is no indication that he found the work particularly fulfilling or that he liked the city itself.[17]

The line that in early 1921 Banting was beside himself with anticipation, common in older, heroic accounts, is pure fiction.[18] He did not even write back to Macleod to discuss coming to Toronto until March and, even after, considered alternative paths right up until 26 April, when he decided that 'nothing [had] presented itself', closed up his practice, and made for the railway station. He did not, however, sell his house in London until months later. Clearly, he felt the situation warranted caution, and thought it wise to keep his options open.[19]

The Early Experiments

When Banting arrived in Toronto, Macleod introduced him to Charles Best, a young student who was to assist with the experiments. The plan seemed simple enough: the two men would operate on laboratory dogs to tie closed their pancreatic ducts. After several weeks, when the organ was sufficiently degenerated, an extract would be made from the remaining tissue. This would then be injected into another dog whose pancreas had previously been fully removed, causing the symptoms of diabetes.

If the theory was correct, and the extract contained the elusive 'internal secretion' with none of the harmful digestive fluid, then, theoretically, the level of sugar in the dog's blood would fall and its condition would improve. The work began on 17 May. Macleod spent several weeks ensuring that his charges understood how to perform the necessary procedures before leaving them to get on with it, departing Toronto in mid-June to spend the summer at home in Scotland.

The experiments did not get off to a particularly auspicious start. While Banting's background in surgery surely helped, neither he nor Best had much experience with this kind of precise technical work. The laboratory space they were given did not help either: it was barely worth the name. The rooms were cramped, dirty, and sweltering in the summer heat, and infections following surgery were common.[20]

Banting was so worried at the rate he was getting through the university's supply of laboratory dogs that he started to covertly replace them with others he found for sale – sometimes in questionable circumstances – on the streets of Toronto.[21] Nineteen dead animals later, on 8 July, he and Best had successfully performed the ligation procedure only twice. Nevertheless, this was enough to test the theory. On 30 July, Banting cut out the shrivelled pancreas of one dog and prepared it according to Macleod's instructions. The process seems macabre by today's standards. The organ was chopped up and mixed with a briny preservative solution, frozen, and then ground into a paste that must have resembled a particularly unappetizing pâté. From here, it was filtered and gradually warmed.

Amazingly, it seemed to work. When it was injected into the dog whose pancreas had been removed in advance, producing diabetic symptoms, the animal's blood sugar fell by over half within the space of an hour. Despite this encouraging start, however, further doses did not appear to be as effective. The next morning, Banting found it desperately sick, and it died a few hours later.

Banting and Best tried again over the following days and weeks, refining their technique as they went. This time, their results seemed much more promising. Injections did often seem to reduce blood sugar. One dog was even brought back from death's door for a brief period, rising from a coma to stand up and wander around. After their last animal died on 19 September, they considered the results over-whelmingly positive, and Banting must have felt that his theory had been vindicated. Even the ever-cautious Macleod, who arrived back in Toronto a few days later, was optimistic, and agreed that further research should be done.

Banting was, however, by this point almost completely out of money. He had not been paid for his work over the summer, and there was simply no way that he could afford to continue. Worse, he had cut all ties with London, selling his house there in early September after becoming convinced that he was onto something big.[22]

During a meeting with Macleod he displayed a new-found confidence, threatening to take his research to another institution if he was not provided with a salary for the winter, someone to help care for the animals, more space to work in, and repairs to the decrepit old laboratory he and Best had been using. The exchange was fiery and it was becoming increasingly obvious that the two men did not like one another at a personal level, but Macleod begrudgingly agreed to his terms.[23]

On 14 November, Banting, Best, and Macleod attended the physiology department's journal club to give a talk on their research. The meeting – coincidentally held on Banting's thirtieth birthday – was a modest affair, and an awkward one for the uncomfortable public

Figure 2.   *Frederick Banting (right) and Charles Best (left) with laboratory dog,*
August 1921
Source: Thomas Fisher Rare Book Library, University of Toronto

speaker. Nevertheless, it prompted one audience member to suggest that a promising next step might be to attempt to show that a dog could be kept alive over a long period of time using the pancreatic extract. All three agreed that this was a good idea.[24]

There was, however, a problem. The extract was slow and difficult to produce. The procedure necessary to ligate the pancreatic duct was complex and still prone to failure. Even when everything went as planned it took weeks for the organ to sufficiently degenerate, and only a small amount of usable material could be harvested from one animal. Securing a consistent enough supply of the extract to keep a dog alive for weeks, months, or longer seemed a tall order.

In the end, Banting came up with a relatively simple workaround. As animals do not need to digest their own food while in the womb, he speculated that foetal pancreas might contain the 'internal secretion' but few or no harmful digestive enzymes. Theoretically, then, effective extracts could be produced from these organs without any complex surgical work, and at a significantly faster pace. When they tested the theory using foetal pancreases sourced from a local abattoir, they found that it worked just as well as the degeneration method.

This solution was resourceful, but it turned out to be unnecessary. In early December Banting and Best had begun to use an alcohol-based preservative solution in place of saline. They used this first with the foetal material and then, to see what would happen, on a non-degenerated adult dog pancreas. Bizarrely, both seemed to work perfectly well. In a later publication, Best reasoned that the use of alcohol as a preservative, made slightly acidic by the addition of hydrochloric acid, neutralized the harmful digestive enzymes.[25] In any case, the extract could now be produced quickly and efficiently from cheap, whole, adult organs available at any common slaughterhouse.[26]

It was now possible to begin long-term testing. Banting and Best were able to keep a depancreatized dog alive from 17 November until 2 December before it abruptly died.[27] Undeterred, from 6 December they tried again. This time the dog – affectionately known as 'Marjorie' – responded well. Receiving daily injections, it survived across the holiday period for a total of over two months. Even more encouragingly, it might have lived far longer – Banting decided to have it euthanized on 27 January. Clearly, he felt the experiment had run its course.[28]

The First Human Trial

Buoyed by his success with 'Marjorie', in early January Banting peti-
tioned Macleod to arrange a human trial. Macleod was sceptical, but he
eventually relented, approaching Duncan Graham, chair of the university's medical department, a senior consultant at the Toronto General
Hospital (TGH), and one of his personal friends. Graham agreed to
have Banting's extract administered to one of his most seriously ill
patients.

There is very little recorded information about the first person with
diabetes to ever receive insulin outside of the context of the TGH. This
is not particularly surprising. Leonard Thompson was a charity patient,
so his family cannot have been wealthy. Amongst the sea of notables
that feature consistently in the early story of insulin, Leonard appears
to have been a remarkably unremarkable young boy.

In 1922, he lay dying. The 14-year-old had started to wet the bed
and complain of tiredness in 1919, and was diagnosed with diabetes
soon after. As was common at the time, he was told to adopt a 'starvation' diet of the kind advocated by physicians like Frederick Allen. He
lost weight over the following years, and by late 1921 was a shadow of
his former self: emaciated, suffering from extreme thirst, and barely
able to summon the energy even to leave his bed.[29]

Leonard was able to secure a place at the TGH only because his
family doctor had connections there and was able to call in some
favours.[30] When he was admitted on 2 December, the staff did not have
high expectations. He weighed barely 65 lb (29.5 kg) and was extremely
ill. By January things were looking dire. Even the most optimistic of
those present doubted he was long for the world. Frank Allan was a
medical student at the TGH when Thompson was brought in, and later
remembered his pitiful state:

> He was admitted to the hospital in a serious condition, emaciated and
> feeble. The dietary regimen employed in the hospital failed to check
> his downward course. He was a pathetic figure as he lay quietly in his
> bed or sat still in the chair at the bedside, too weak to show interest
> in the activities of the large, busy ward. All of us knew that he was
> doomed.[31]

That Thompson had been able to secure a place at the TGH at all was fortunate for a child of his station. Yet, despite – or, perhaps more cynically, precisely because of – his socio-economic situation, he was chosen to be the first to receive Banting and Best's experimental extract.

In a series of events that would give the members of most twenty-first century ethics committees heart palpitations, on 11 January 1922 Leonard was injected with what one of those present later described as 'thick brown muck'.[32] It was not a success. While it did slightly reduce the boy's blood sugar, it did nothing for his overall condition and he developed a painful abscess. No more of the extract was given. Banting tried to put a positive spin on things – his blood sugar had fallen, after all – but there is little doubt that he was dismayed.[33]

The first human trial of the pancreatic extract had fallen far short of expectations. The experiment had clearly been premature. Animal studies were still ongoing, and while they had worked continuously to tinker with and refine the production process, Banting and Best were still unable to reliably produce anything with a standardized potency. Some batches caused a dramatic reduction in blood sugar while others appeared to have almost no effect at all, and side-effects remained troublingly unpredictable.

Only a month before the trial with Thompson, however, Macleod had – at Banting's request – enlisted the aid of James Collip. Collip was not a medical doctor but an experienced and well-respected academic biochemist in Toronto on sabbatical from his usual post at the University of Alberta. Throughout December, his considerable technical skill allowed him to gradually produce a far more consistently effective pancreatic extract than Banting and Best could have hoped to achieve.

Collip was not at all happy when he discovered that Banting's extract had been tested on Thompson. It had been made entirely behind his back, and with none of the innovative techniques that he had developed. He later described it as 'absolutely useless for continued administration to the human subject'.[34]

Why had Banting acted so recklessly? The most likely answer is that he was simply insecure, and jealously protective of a project he considered very much his own brainchild. He had always been concerned that Macleod would attempt to steal credit for his idea – this was one of the key factors behind their strained relationship. When Collip joined the team and made vast improvements to the extract production process over a relatively short period of time, it highlighted just how underqualified both he and Best actually were.

If Macleod had chosen to marginalize them, and the first human trial had been performed with an extract they had no hand in making, then they might easily have been written out of the history – remembered only as junior collaborators who worked alongside the real innovators – Macleod and Collip. This was not an entirely unreasonable fear on Banting's part. By early 1922 he and Best had reluctantly taken more of a back seat. The priority had become producing the extract in as pure a form as possible, and this was work that neither were equipped to meaningfully contribute to.

Banting had staked everything on the project, and he found the idea that he might be denied credit utterly humiliating. Everything that had been achieved, he felt, originated with *his* idea, scrawled on a notepad by his bed over a year prior in London. By pushing for an initial clinical test on 11 January with his and Best's extract, Banting placed himself into a firm position of precedence. If the experiment worked, he could celebrate. If it did not, he had still performed the first human trial of the research group's extract, and any later improvements would be read as refinements to *his* invention.

Ironically, it is usually Collip who is forgotten in the story of insulin – in no small part thanks to Banting and Best's conscious choice to undermine his importance in later years.[35] In reality, he did more than anyone to transform pancreatic extracts into a viable clinical treatment for diabetes. Thompson certainly owed him his life. After Banting's ill-advised trial, Collip worked alone in the university's pathology department day after day and often long into the night, performing countless experiments in an effort to refine the 'active principle' of the extract – the precise chemical part of it that reduced blood sugar, free from other contaminants. By 19 January, he thought he had succeeded. A few days later, on the 23rd, Thompson was given another injection using Collip's 'sterile and highly potent' formula.[36] It had immediate and dramatic effects. His blood sugar fell markedly and his physical condition visibly improved. Against all the odds, it looked like he was going to make it.

The Reaction

The Toronto group excitedly produced a flurry of articles in February 1922. Throughout, they were careful to manage expectations. In the most significant – a piece for the *Journal of Laboratory and Clinical Medicine* – Banting and Best described their animal experiments throughout late 1921 and early 1922. While clearly enthusiastic about the potential of their results, they warned against inferring that the new extract could be used to successfully treat humans with diabetes.[37] Their caution was well-justified – Thompson's first injection on 11 January had not produced the immediate recovery they had hoped for, and while he was considerably improved after being given Collip's purified extract later in the month, there was still much uncertainty about its long-term effectiveness.

By March, however, they were more confident. In an article for the *Canadian Medical Association Journal* that focused primarily on the trials being conducted on Thompson, they concluded optimistically. By using the extract, they claimed, it was possible to effectively neutralize the symptoms of diabetes:

(1) Blood sugar can be markedly reduced even to the normal values.
(2) Glycosuria can be abolished.
(3) The acetone bodies [ketones] can be made to disappear from the urine.
(4) The respiratory quotient shows evidence of increased utilization of carbohydrates.
(5) A definite improvement is observed in the general condition of these patients and in addition the patients themselves report a subjective sense of well being and increased vigor for a period following the administration of these preparations.[38]

At the beginning of May, Macleod travelled to Washington D.C. to present a paper to the conference of the Association of American Physicians (AAP). This was, and remains, an important annual event, and numerous luminaries from across the medical profession were present, including Allen – widely considered the world's leading authority on diabetes. While the team's research had received some prior attention, mostly from the local press, this was the talk that really put a spotlight on Toronto.

Macleod's presentation was received with overwhelming enthusiasm.[39] Elliott Joslin recalled that a 'standing vote of thanks' was moved to 'express to Professor Macleod and his co-workers . . . appreciation of his epoch-making discovery'.[40] Even Allen was mightily impressed, remarking that insulin, as the Toronto group were now calling it, might well be 'one of the greatest achievements of modern medicine'.[41] The despair associated with those desperate, 'severe' cases of diabetes that seemed universally terminal might, it seemed, yet become a thing of the past.

By May, Thompson was considered well enough to return home, and he was discharged on the 15th.[42] Allan, who had seen his condition when he was first admitted, later recalled crossing paths with the now adolescent boy as an intern years later, and being astounded at his apparent health:

> Three years later, when I was an intern in the Toronto General Hospital, I saw Leonard Thompson when he came in regularly to secure his supply of insulin. He was now a sturdy young man, who showed little resemblance to the emaciated, dying boy who had been the subject of the most crucial clinical experiment in the field of diabetes.[43]

In all of the fanfare that followed, it is easy to forget that the first human to benefit from insulin disappeared into relative obscurity, living a quiet life as a factory worker until his death from pneumonia only thirteen years later. While Elizabeth Hughes became – much to her own discomfort – something of a poster-child for insulin, Thompson is rarely even referred to by name in contemporary reporting, and we know next to nothing about his later years.[44]

Similarly, it is difficult to imagine someone like Hughes receiving that first experimental injection. That it was Thompson who was chosen was telling. What little privilege he and his family had was sufficient to have him admitted to the TGH as a charitable patient. Ironically, however, on the ward his humble status set him aside from his fee-paying equivalents. The subtext here is clear: Thompson was an unimportant boy from a working-class background, used as a human guinea pig and then cast aside and ignored in favour of his more glamorous contemporaries when his expendable body was no longer required.[45] In this grand story, he is portrayed as a passive object rather than a unique person in his own right. As if to hammer home the point, after his death his pancreas was removed and kept as a trophy, placed on display at Toronto's Banting Institute.[46]

The Aftermath

The move to large-scale insulin production came with some significant teething problems. At one point in early 1922, for example, Collip simply forgot how to make his purified extract. Despite his brilliance as a chemist, he was, apparently, not the most organized of characters, and had neither kept detailed notes nor told anyone else the exact process. What resulted was a desperate scramble to rediscover an effective method as vital supplies dwindled, forcing the group to ration what they had available. Thankfully, by mid-May, they had achieved a much-needed, albeit embarrassing, second breakthrough, and production could finally begin on an industrial scale.[47]

From the outset, the Toronto group set out to ensure that insulin would be made available to as many people as possible, regardless of wealth. Despite reservations about patenting their discovery, they were concerned that someone else – smelling profit – might develop a sufficiently modified version and choose to do so independently, potentially snatching the practical manufacturing rights from under their noses. Understandably, they quickly decided to begin the process of applying for their own. Macleod later made it clear that this was a defensive decision, reflecting in 1924 that the primary objective had been 'preventing any other person from taking out a similar patent which might restrict the preparation of insulin'.[48] When the patent was granted to Banting, Best, and Collip in January 1923, they immediately transferred the rights to the university's board of governors for a dollar apiece.[49]

The Connaught Anti-Toxin Laboratory just outside of Toronto had been contracted to begin production of insulin in February 1922, but this small, university-affiliated institution suffered major technical issues from the outset, and even when things were operating as expected could only make so much. By May it was increasingly obvious that production needed to be scaled up considerably if it was going to meet demand.

For some time, George Clowes, a representative of then-modest Indianapolis drug manufacturer Eli Lilly and Company, had been making it very clear that his employers were extremely keen to get in on the insulin game. The firm's leadership were (correctly) convinced that the new 'miracle drug' could be highly profitable in the long term, and might establish them as major players in the pharmaceutical market. They did not, of course, say as much in discussions with Toronto. Instead, they tugged at the Canadians' heartstrings. There was simply not enough insulin to go around, and most of what was produced went to the very wealthy. Was this not unfair?

Toronto's board of governors were hesitant, but it was true that demand for their new lifesaving medication was significantly outstripping Connaught's supply capacity. They came to an agreement with Eli Lilly at the end of May. The company would have the exclusive right to produce and distribute insulin throughout most of the Americas, Canada excluded, for one year. In return, during this period, they would be expected to share any improvements to the production process, alongside any new patents that resulted for markets outside of the United States, where they would retain sole rights. The Americans would also send free shipments of insulin to Toronto and a small group of selected physicians and institutions for experimental purposes, and they would sell the surplus at cost once production was sufficient. Eli Lilly clearly expected to be given a more robust long-term licence after the initial agreement expired, but they demanded no further exclusivity.[50]

On the surface this seems like a very generous deal uncharacteristic of the private sector, but Eli Lilly obviously had a long-term plan. By the terms of the deal, they would learn how to make insulin before any of their competitors, and the agreement all but guaranteed a further contract with better terms at a later date. Being able to retain single ownership of the US patents for any new innovations could potentially reap great rewards, but even if they did not make any technical advances, this first year was a valuable head start. Eli Lilly could perfect their operation before anyone else had a chance to begin, while also generating name recognition for their branded version of insulin: 'Iletin'. Even better, they could do all of this while projecting the image of an 'ethical' company interested in more than just shrewd business. Nonetheless, potential profit was always a top priority.

In a rather revealing moment in 1923, the company's president J.K. Lilly even pushed Clowes to make the case that Toronto should renew the exclusivity contract indefinitely. It had, he said, become 'proper for us to carefully consider the commercial phase', arguing that 'best results' could be achieved by 'licensing but one maker in each of the principal countries of the world'.[51] This particular demand was refused, but it is clear that the deal was always considered an investment first and foremost, and one that would ensure a significant edge in the years to come.[52]

Insulin production in North America remained inconsistent until the last few months of 1922, by which time Eli Lilly had effectively ironed most of the kinks out of their manufacturing process. By the early months of the following year, they even built up a surplus, and they were able to make up for a continued shortfall in Canada, where the Connaught Laboratory's capacity still lagged, while also beginning to ship supplies further afield.

Across the Atlantic and beyond, news of insulin had been met with considerable initial scepticism. The idea that a simple injection could effectively end the suffering of so many desperately ill people seemed far too good to be true, and European doctors treated the supposed 'miracle cure' with suspicion. As time rolled on, however, it became impossible to ignore that something of great importance had been achieved in Canada, and many began to loudly demand that it be made available for their own practices.[53]

From the second half of 1922, Toronto began to license insulin production internationally, eventually extending rights to twenty-five countries across the globe.[54] By early 1924, supplies were (at least in the wealthier countries that were considered important) secure. Eli Lilly continued to be the biggest manufacturer, but Connaught had managed to ramp up production sufficiently enough that Canada no longer relied on imports from the United States. Despite a slow start, European operations in Britain, Denmark, and Poland were also well underway.

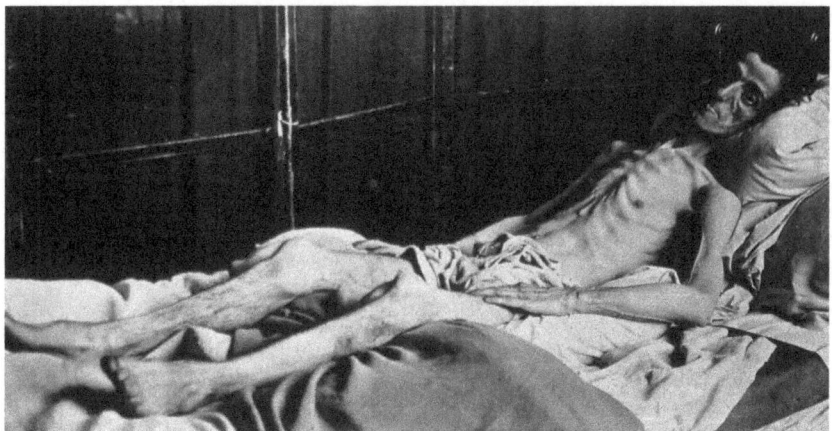

Figure 3.   *Before insulin treatment, c.1925*
Source: Wellcome Collection

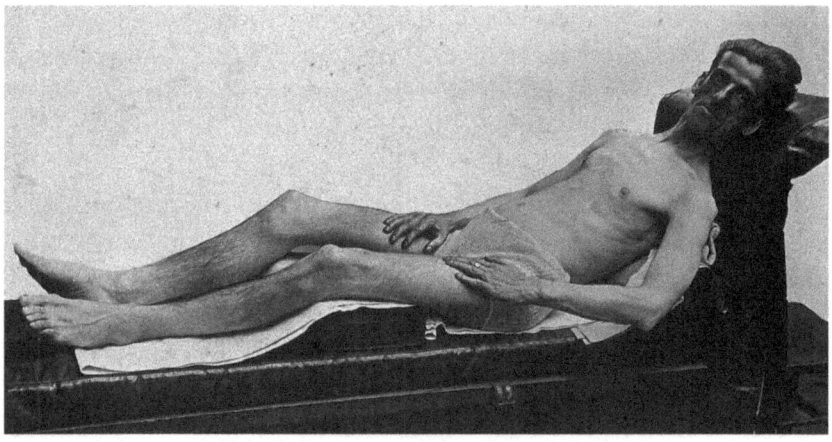

Figure 4.   *After insulin treatment, c.1925*
Source: Wellcome Collection

Those who saw insulin in action were almost universally amazed, and with good reason. Before-and-after photographs from the period retain a sense of genuine wonder (see Figures 3 and 4). Their initially emaciated subjects – many of them children – are shown restored to health, seemingly miraculously, within a matter of days or weeks. Initial caution notwithstanding, by mid-1923, few could deny that this was something ground-breaking.

Banting was propelled to fame almost overnight. He was made an attending physician at the TGH in August.[55] This was a significant promotion with a salary to match, but it was probably granted more out of concern that he might jump ship and take his now considerable reputation elsewhere than because of his great skill as a clinician. It was around this period that he saw Hughes and others – most of them from a very different world to Thompson, the first successful insulin patient.

By the autumn of 1923, he had become a household name, and received a variety of honours. The most impressive, perhaps, came in June, when the Canadian parliament voted to award him an annual stipend of C$7,500 – over C$100,000 in today's money – in recognition of his achievement.[56] Within the space of two years, he had risen from obscurity to become one of the country's most celebrated figures. This, perhaps, reinforced a certain sense of self-importance. In October 1923, Banting learned that he was to be awarded the Nobel Prize for his work – the first ever to go to a Canadian. It should go without saying that this is one of the greatest marks of public recognition that any scientist or physician can hope for. The vast majority – even the most talented – will never come close to achieving it.

With this in mind, Banting's reaction seems incredible. He had won the prize, but it was to be shared with Macleod. Relations amongst the researchers had always been strained, and tension about how to divide credit had been bubbling under the surface since at least early 1922, but the Nobel announcement caused it to explode into outright hostility. Banting was incandescent with rage at the prospect of Macleod getting half of the credit while Best – who he felt had contributed significantly more to the project – was sidelined. He clearly felt that all of his anxieties about his work being stolen were being proved justified. Furious, he even declared that he planned to reject the award entirely. In the end he did not, and the ceremony went ahead, but he insisted on sharing his half of the prize money with Best. Perhaps backed into a corner, Macleod, after some deliberation, split his with Collip.

For what should have been a triumphal celebration, the mood was sour, and the men split on very poor terms. Macleod and Banting's relationship in particular was in tatters, and the younger man held a lifelong grudge. They went to their graves having never reconciled – Macleod in his hometown of Aberdeen in 1935 after several years of ill-health, and Banting, rather more dramatically, following an aeroplane crash in Newfoundland in 1941, en route to Britain during the Second World War.

Evaluating Toronto

So, a century later, how should we look back on the 'discovery' of insulin? The story is impressive, but the traditional narrative tends to miss some important details. To start, Banting was by no means the first person to create a pancreatic extract that could successfully reduce blood sugar in either laboratory animals or humans. This had actually been achieved several times since Minkowski and von Mering first drew attention to the hypothetical 'internal secretion' in 1889.

The most prominent pre-Banting extract was probably that made by Berlin's Georg Ludwig Zülzer. In 1906, Zülzer injected a comatose man likely experiencing DKA with a pancreatic extract he called 'acomatol'. While he remained very ill, the man's symptoms significantly improved for a brief period of time before he relapsed and died. Zülzer managed to secure funding for further trials, but the results were mixed. 'Acomatol' did often relieve the worst symptoms of diabetes, but it was inconsistent, and those treated frequently experienced debilitating side-effects – sometimes even violent seizures.[57] Zülzer remained optimistic, but his failure to solve this problem dissuaded the universities and pharmaceutical companies that might have supported him, and by the 1910s he was increasingly marginalized. In 1914, the First World War broke out and he was conscripted, putting an end to his research. In the years following the conflict, war-scarred Germany was no place for a desperate, middle-aged scientist lacking credibility, and a Jewish one at that. Mercifully, Zülzer was able to escape the Nazi regime. He emigrated to New York in 1934, where he faded into relative obscurity.

To his credit, Banting never claimed precedence in developing *any* effective pancreatic extract. In fact, by the late 1920s he openly acknowledged Zülzer, writing that that very first experimental injection that had been given to Thompson on 11 January 1922 had shown no meaningful improvement over the German's 'acomatol'.[58] Instead, he believed that his contribution was to produce something that consistently worked while producing few side-effects. It was a shame that the extract used in that first test appeared to have been from a dud batch, but more often than not it *did* seem to function as intended – he and Best had, after all, kept a dog with no pancreas alive for months. From Banting's perspective, all that was necessary was a little refinement, which Collip subsequently achieved. The key idea, though, had been his: to protect the 'internal secretion' from the destructive digestive enzymes.

Once again it is important to stress that Banting was not in any way a specialist in the subject of carbohydrate metabolism. At best, he had what cursory knowledge he had picked up during his rushed medical education, and he had probably learned a little more during his brief time working at Western. Had he dug further into the literature, however, he might have discovered that his 'eureka' moment – ligating the ducts so the pancreas degenerated, then harvesting the 'internal secretion' from the remaining islet cells – was not quite the unique insight that he thought it was. In fact, the University of Michigan's Lydia DeWitt had pre-empted him by almost a decade and a half, using the same process on cats in 1906. Her extract seemed to have some effect in test-tube experiments and she speculated that such a preparation might one day be used in treating diabetes, but she performed no studies on living creatures.[59]

Banting could never have known it, but there is also evidence that Frenchman Eugène Gley came up with the same theory even before DeWitt. In December 1922, when Toronto was making headlines around the world, Gley revealed that he had left an envelope in the care of the Society of Biology in Paris all the way back in 1905.[60] When it was opened, it was found to contain details of turn-of-the-century experiments in which he had successfully used an extract made from degenerated pancreas to reduce the blood sugar of laboratory dogs. Gley had abandoned the research for lack of resources and, inexplicably, had never bothered to publish on the subject.

In any case, why does this matter? Maybe others *had* previously had the idea to preserve the 'internal secretion' of the pancreas, by ligating the ducts or otherwise. Banting's achievement was not that he was the first to come up with the theory, but the first to take the idea far enough to produce insulin – an extract preparation that could reliably be used clinically to treat diabetes.

Yet here is probably the greatest irony of all. While in theory a good idea, his method was completely unnecessary. It was actually the process of immediately chilling the extract that prevented the deterioration of the 'internal secretion'.[61] The digestive enzymes had always been mostly harmless, as their destructive power is not activated until they reach the small intestine. Banting's great contribution was, in hindsight, totally irrelevant. Quite how he failed to realize this is genuinely bizarre. As early as August 1921, before Macleod had even returned to Toronto from Scotland and long before Collip joined the team, he and Best had tested an extract made from whole, non-degenerated pancreas and had seen quite clearly that it worked, but they downplayed what should have been a remarkable result.[62] Despite Best's later speculation about the implications of switching to alcohol as a preservative, Banting never seemed to reflect on the deeper implications of moving to whole-organ extracts for his theory.

It would probably be overly cynical to read any deliberate bad faith into this. Banting was not equipped to engage in extensive self-criticism in this area even if he had wanted to do so – he was no expert in the theoretical principles at work and never became one. Nonetheless, the truth is that many others had previously done much of the same work as the Toronto group and had developed essentially viable pancreatic extracts but had been kept from success by sheer misfortune.

Gley, mentioned above, probably falls in this category, but his decision to remain silent about his work was a strange one. Israel Kleiner, too, could easily have been celebrated as the 'discoverer' of insulin had it not been for circumstances entirely beyond his control. In 1919, while working at New York's Rockefeller Institute in the laboratory of Samuel James Meltzer, Kleiner published a paper that described how, in 1915, he had recorded excellent results using an extract of whole pancreas on laboratory dogs. The disruption of the First World War had delayed his work, but he had managed to replicate the results. Unlike Zülzer, he reported no serious side-effects beyond an occasional mild fever, and he outlined a relatively simple manufacturing process.

Kleiner clearly understood the clinical implications of his discovery, writing that 'the temporary effect which it produced in dogs might be duplicated in man', but he remained cautious, warning that 'it is doubtful whether attempts along this line should be made until further knowledge has been obtained.'[63] Unfortunately, before he was able to conduct any human tests, Meltzer retired – ironically enough following a diagnosis of diabetes. The laboratory was shuttered and Kleiner was made redundant. While he quickly found another post at the New York Homeopathic Medical College and went on to have a successful career, he never revisited the question of pancreatic extracts.[64]

Perhaps the most devastating precursor to Banting, however, was Ernest Lyman Scott. Scott, a postgraduate student at the University of Chicago, had been inspired to investigate pancreatic extracts to honour a friend who had died of diabetes. In 1911, he submitted a thesis in which he claimed to have effectively suppressed diabetic symptoms in dogs, concluding – quite correctly – that:

1. There is an internal secretion from the pancreas controlling sugar metabolism.
2. By proper methods this secretion may be extracted and will retain its activity.[65]

Scott's experience will sound very familiar. He had the same idea as Banting – to extract the 'internal secretion' by the ligation method. In his case, he quickly gave up on this strategy because the procedure was complex and he was no surgeon. Instead, he adopted a different approach involving alcohol and rapid cooling, once again just as Banting and Best had in late 1921. This appears to have worked well.[66] Scott certainly would have liked to have continued his work – he was probably more qualified than the Canadian, all things considered. However, his research received a lukewarm reception. Years of failure had led many experienced researchers in the field to believe that pancreatic extracts were a busted flush, and he was dismissed out of hand.

The cruellest irony of all came when Scott travelled to Cleveland's Western Reserve University in 1913 to discuss potential further research with a leading authority on carbohydrate metabolism. That person, as it turned out, was John Macleod. When they met, Macleod was polite enough but remained unconvinced. He agreed that the 'internal secretion' of the pancreas probably existed, but did not at that time believe that it could be effectively isolated, and he doubted whether Scott's results were repeatable. Without institutional support, Scott dejectedly abandoned the endeavour and moved on. He was in the audience when Macleod gave his talk to the AAP conference in 1922, and seems to have been suitably nonplussed. When he found out about the 1923 Nobel Prize, he did cordially congratulate the recipients, but one can only imagine how he must have felt.[67]

Perhaps the loudest protests came from Romania, where Bucharest's Nicolae Paulescu, a deeply unpleasant man whose enthusiasm for medical research was surpassed only by his love of fascism, angrily claimed that he had successfully used a pancreatic extract on laboratory animals in 1919 and should have precedence. In his defence, he had published papers to this effect in 1921 (which Banting was aware of), and in early 1922 conducted some provisional human trials, albeit with mixed results.[68] Paulescu's extract, 'pancréine', was, much like Zülzer's 'acomatol' not reliable enough to use clinically. The Nobel Prize committee was unmoved.[69]

That said, *Banting*'s extract was not reliable enough to be used clinically either. His apparent stroke of genius – ligating the pancreatic ducts, waiting for the organ to degenerate, and harvesting the 'internal secretion' from what remained – was not an original idea, and in any case played no role in making his preparation effective. His and Best's extracts were no more consistently useful than many of those that had come before. If anything, it was Collip who was responsible for the great breakthrough that made insulin clinically viable.

Is the mythology around Banting – still one of Canada's national heroes – warranted by his scientific contributions? Probably not. His 'great idea' was unoriginal and ultimately incorrect, and the experiments he conducted with Best over the summer of 1921 were sloppy.[70] Their results seemed impressive at the time, and for a 'first, unaided attempt at research by two young enthusiasts', they were, but Banting and Best were nonetheless amateurs covering old ground on a false premise.[71]

After receiving the Nobel Prize, Banting had every resource he could imagine at his disposal, but he never came close to replicating the success of insulin. The jealous insecurity he displayed throughout his later life, and in particular his hostility towards Macleod, suggests that at some level he understood how fragile the foundations of his prestige actually were. Nonetheless, he *was* instrumental in the discovery of insulin as a pharmaceutical product. His contribution was ultimately not scientific, but to act as the indefatigable driving force that brought together the right people at the right time and persuaded them to believe in the idea of a workable pancreatic extract – something that many in preceding years had written off as fantastical.

Insulin was a collective achievement. Banting's sheer willpower was instrumental in the discovery, but so was Best's tireless assistance and Collip's technical expertise. Even Macleod, who is often cast as an admittedly entertaining villainous caricature in retellings of the story, played an essential role by providing the initial support and, later, using his connections to spread the word.

If the story of insulin's 'discovery' teaches us anything, it should be that scientific innovation is almost never the work of one person. This reality is often blurred by awards like the Nobel Prize, which cannot be formally shared between more than three recipients.

In the early twentieth century, there were countless potential Bantings. He, like the rest of them, might well have failed had things been only slightly different. In the event, it seems a miracle that he did enjoy the success that he did. Like Scott, he approached the esteemed Macleod to discuss his idea, and, like Scott, left the professor unimpressed. But unlike Scott, Banting was given a chance.[72] We will never know precisely why, but the lives of thousands were changed forever as a result.

Chapter 2

Insulin in Practice, 1922–1978

When it was introduced to the world in 1922, insulin seemed a miraculous triumph of medical science. The reception from the press was predictably uproarious. 'Insulin is Cure for Great Scourge' read one headline, 'Diabetes, Dreaded Disease, Yields to New Gland Cure', another. 'Diabetes is Robbed of Its Menace'. 'Diabetes Is Completely Checked by Insulin'. The mood was jubilant.[1] For a time, there were hopes that insulin would amount to an absolute cure. If diabetes got worse because excess sugar in the blood overworked the islet cells, as most physicians of the time believed, then it stood to reason that a short course of injections might break the cycle by taking pressure off them entirely and allowing them to recover.

In a January 1923 lecture in Detroit, John Macleod optimistically suggested that 'given a complete rest by the administration of insulin, the pancreas will function normally again without further treatment'.[2] A few months later, Frederick Banting wrote in an article for the *Journal of Metabolic Research* that it 'seemed possible that the degree of rest made possible by the use of insulin might in certain cases bring about sufficient increased carbohydrate tolerance as to make it possible to dispense with its continual administration'.[3]

The idea of 'resting' the pancreas was based on an understandable but erroneous early twentieth-century understanding of what diabetes actually was. T1DM, which probably accounted for almost all 'severe' cases in the 1920s, is autoimmune in origin. The destruction of the islet cells has no relation to blood sugar levels. T2DM, on the other hand, is not even primarily pancreatic in origin – it is the product of insulin resistance.

Despite contemporary hopes, wherever insulin therapy was suspended, symptoms quickly returned. Leonard Thompson, for example, was sent home from the TGH in May 1922 without any. He only obtained a repeat prescription after he was readmitted in October with DKA. In July 1923, William McCann of Baltimore's Johns Hopkins University Hospital reiterated that any claims of an absolute cure in the press were vastly exaggerated:

> The most that can be said for insulin is that it is a specific remedy for diabetes which restores the metabolism to normal as long as the treatment is continued. In some cases the beneficial effects may continue for a short period after discontinuing the treatment. Sooner or later the patients always return to the condition preceding the treatment unless it is resumed.[4]

Nonetheless, even if it did not technically qualify as a wholesale cure, insulin surely appeared a functional one in practice, particularly to those looking on from a distance. In April 1923, Britain's *Westminster Gazette* wrote triumphantly of its potential to grant new life to those who otherwise would almost certainly have perished, and to render their condition essentially only a minor inconvenience:

> Cases which a few years ago would have been the despair of the doctor are now providing instances of cures that look almost miraculous . . . The insulin has apparently to be used permanently, but . . . the sufferer who would once have been condemned to wastage and death need now take no further care of himself than is imposed on thousands of people for one reason or another.[5]

The *Gazette*'s sentiments were characteristic of the time and clear in their implication: insulin was not an absolute cure, but it might as well be. While some reference was occasionally made to the need for continued dietary regulation, this was usually skimmed over quickly.

The overall tone cast insulin as a magic bullet. Diabetes had, apparently, been conquered.

The Early Insulin Era

Any claim that insulin trivialized diabetes was, of course, a dramatic exaggeration. It could reliably bring blood sugar down and relieve symptoms, certainly, but long-term use posed significant challenges. How much, for example, should be injected, and when? Most physicians in the 1920s agreed that the primary goal of treatment should be to keep the urine clear of sugar, but safely accomplishing this was trickier than it first appeared.

James Collip had noticed as early as January 1922 that, given enough insulin, rabbits would become hypoglycaemic. They would get confused and irritable, and, with increasing doses, would even lose consciousness or die outright. Determining precisely how much was sufficient to cause convulsions in these animals actually ended up becoming a useful strategy for standardizing its potency.[6]

Similar reactions were quickly seen in the first cohort of human insulin recipients. Discussing their initial clinical experiences in the *BMJ* in early January 1923, the Toronto group pointed out that while hypoglycaemia did seem relatively easy to remedy with a little fast-acting sugar, it nonetheless had the potential to be quite dangerous:

> In giving a dose, therefore, to render the patient sugar-free it sometimes happens that the blood sugar falls well below the normal level, and this sudden hypoglycaemia is accompanied by a characteristic train of symptoms . . .
>
> These reactions can be relieved by food administration; 50 to 100 c.cm of orange juice has an almost immediate effect in clearing up the symptoms . . .
>
> Up to the present time no serious mishap has occurred as a result of these hypoglycaemic reactions, but while this is so it is felt that hypoglycaemia constitutes a real source of danger.[7]

In short, treatment promised to be complex. It quickly became obvious that maintaining consistent blood sugar levels low enough to ensure that the urine remained free of sugar but high enough to avert the risk of hypoglycaemia was no straightforward task. Even more challenging, daily insulin requirements are never static. Carbohydrate intake is the most obvious source of fluctuations, but exercise, sickness, stress, and other factors can all play their part. The functional pancreas dynamically varies the amount of insulin that it produces as needed, but this is impossible where it is delivered manually. A syringe holds only as much as it does, and delivers its contents only occasionally.

To compound this difficulty, only one kind of insulin was available in the 1920s: the same kind developed in Toronto.[8] Soluble – or, in North America, 'regular' – insulin acts on blood sugar relatively quickly and lasts for a total of around eight hours.[9] It does not, however, do so evenly. Beginning slowly, its activity peaks at, on average, between two and three hours post-injection, after which it quickly begins to taper off. This is a relatively long and inflexible pattern, and, if at any point demand for insulin does not match what is available, blood sugar will rise or fall accordingly.

So how did contemporary physicians respond? After he was through warning that insulin was not a cure, McCann went on to chastise the press for its 'careless' overselling of the new treatment. Those who benefited from it, he maintained, were not freed from any deprivation, and dietary control remained paramount:

> Accurate control of the diet is more necessary with insulin than without it. The reason for this is twofold. A given dose of insulin will cause the proper utilisation of a fairly definite amount of food. If more food is taken than the dose of insulin provides for, the patient will have sugar in the urine again, so that some of the good effects of the treatment will have been nullified. On the other hand, an overdose of insulin may kill the patient by reducing the sugar of the blood below that which is necessary for life. This means that the diet of the patient must be measured as to insure that there will be the right amount of food taken to balance the dose of insulin given. The patient who takes insulin is given a liberal diet, but the diet must be accurately measured, and the patient must take all that is prescribed.[10]

Where McCann talked about a 'liberal' diet, he did not mean to suggest that his patients were encouraged to eat freely, only highlight that they no longer needed to endure punishing 'starvation' treatment. Frederick Allen agreed, and pointed out that significant 'disillusionment' might occur if expectations were not moderated. Dietary prescription could be relaxed, he insisted, but never abandoned.[11]

While physicians often had their own idiosyncratic approaches to treatment, in the 1920s almost all of them attempted to ensure that their patients maintained stable blood sugar levels by emphasizing the need for disciplined adherence to a fixed pattern of diet and insulin, and more generalized guidelines around lifestyle that tended to emphasize moderation. This, however, posed its own problems. Insulin therapy was a new frontier for medicine because it had to be conducted away from professional oversight in almost all cases. Only the very wealthiest could afford to have a doctor or nurse attend them every day. Furthermore, it was not like most other treatments – it was significantly more complex than taking a pill every morning, for example.

Management outcomes might change dramatically based on small changes to dosage, injection timing, or diet. Furthermore, all of these were, in practice, controlled absolutely by the layperson. In addition to choosing whether or not to implement treatment, the material context of insulin therapy put them in a position in which they could engage with and shape it at will, whatever their doctor said. This could cause no small amount of anxiety amongst professionals. For some, this stemmed from a genuine concern for the welfare of their patients. F.G. Brigham, a physician in Boston, for example, claimed that, only months after the introduction of insulin, three people had already died by accidental overdose. The substance, he claimed, was 'a dangerous drug when used by anyone but a qualified doctor'.[12] Implicitly, however, insulin also worked to undermine the physician as the final authority on decision making in treatment.

If they were to treat themselves at home, it was considered imperative that people using it understood exactly how to do so – or rather, exactly how their physician expected they do. Guidebooks aimed directly at laypeople were one common method of achieving this, and several authority figures produced publications of that nature during the 1920s and 1930s – Elliott Joslin's *A Diabetic Manual for the Mutual*

Use of Doctor and Patient, for example, being perhaps the most famous.[13] Almost all of these emphasized the need for quiet moderation, passivity, and obedience on the part of their readers, and contained the implicit suggestion that, should they fail to adopt the attitude expected of them, they would be responsible for any subsequent misfortune. In doing so they worked, as Chris Feudtner argues, to morally indoctrinate people with diabetes with 'values of intensive regulatory control'.[14]

Joslin's British counterpart was R.D. Lawrence, who, in 1934, co-founded the British Diabetic Association (BDA) alongside novelist H.G. Wells.[15] Lawrence was a fascinating character, and, while he promoted similar values, was perhaps one of the most empathetic contemporary practitioners towards his patients. Part of the reason for this was that he was one of a vanishingly small proportion of physicians who were personally reliant on insulin even while they prescribed it, and he well understood the trials of living with diabetes. Lawrence had been diagnosed as a surgeon at King's College London in 1920, and, certain that his condition was terminal, soon moved to a less taxing post in Florence. He almost certainly intended to die in this pleasant Italian environment, and hoped that the distance might spare his loved ones the pain of seeing his deterioration. In the event, however, he got lucky.

In May 1923, Lawrence received a telegram from his colleague Geoffrey Harrison in London. The message was brief and to the point: 'I've got some insulin. Come back quick. It works.' Arriving back in Britain at the end of the month after a near-thousand mile race across a Europe still scarred by the First World War, his life was saved, and he dedicated the rest of his career to treating the condition that had so nearly killed him. He went on to open one of Britain's first specialist diabetes clinics at King's and, before long, had become one of the country's foremost authorities on the subject.[16]

Nonetheless, Lawrence's approach also reflected the paternalistic ideology of the early twentieth century. He was committed to the enforcement of a particular set of rules and attitudes, arguing in the 1925 first edition of his handbook *The Diabetic Life* that 'the training of the diabetic in his life-long creed is the most important part of his treatment'.[17] *The Diabetic Life* comprises advice covering almost every facet of life with diabetes. Lawrence provides practical instructions on injection technique, testing urine for sugar, diet, the management of hypoglycaemia, and similar topics. Most tellingly, however, he also provides strict guidance about lifestyle. Eating out and taking holidays, for example, were discouraging for all but those he considered the most 'skilled and self-controlled'.[18] Even visiting friends, he thought, could be problematic. Where someone with diabetes 'cannot refuse [food] without offence', for example, Lawrence insisted that, they 'must never accept another invitation to that house.'[19]

Ironically, Lawrence's insight into the texture of life with diabetes provided him with the knowledge and tools through which to ever more comprehensively and persuasively promote the 'diabetic creed'.[20] In reality, however, he followed few of his own rules with any real consistency. He travelled and ate out frequently, smoked cigars and drank brandy, and was on at least one occasion witnessed injecting his insulin directly through his trousers in a restaurant toilet rather than carefully preparing the syringe at home as he expected those under his care to.[21] The 'diabetic creed' was, after all, for patients, not doctors.

Sometimes, this moralized 'education' was made obligatory. In order to prepare people for life using insulin, they were often expected to spend a period of time under the close supervision of healthcare professionals. In *The Diabetic Life*, Lawrence alluded to this when he wrote that 'at first the patient must be under close observation'. Before this could be dialled back, he insisted, the doctor should be satisfied that they could 'carry out the following points':

1. He must be able to work out and follow the prescribed diet . . .
2. He must be able to examine his urine by Benedict's test for sugar.
3. He must, if it is necessary, be able to give insulin skilfully and accurately himself, or have some one on whom he can *regularly* depend for this. The correct relation between the time of injection and meals must be observed.
4. He must know the symptoms of an overdose of insulin (hypoglycæmia) and the remedy – rest, probably one or two lumps or sugar or a small orange.
5. He must never break the proper balance of diet and insulin . . .
6. If ill, or if acidosis is shown by a positive ferric chloride test, he must go to bed and inform his doctor.

7. He should visit his doctor regularly, taking with him notes of his diet, weight, and the results of his tests, and of any questions he wants to ask.[22]

This time under 'observation' was usually – though not always – spent in a hospital or other facility where the new patient could be monitored and controlled, and was generally referred to as a period of 'stabilization'. During this time, an appropriate daily prescription of diet and insulin would be established and, just as important, 'education' would be instilled into the person admitted. The duration of this process could vary. Joslin felt that 'intelligent patients can be taught the use of the diet and insulin in a week, and in two weeks the average patient can become free of acid and sugar, [and] learn what is requisite either in hospital or in boarding house, or with a diabetically trained nurse in his own home.'[23] Lawrence thought that some would meet the necessary standard within 'a few hours' and others a 'few days or weeks', while a small number would always struggle and would 'have to depend on a capable relative or nurse for the rest of their lives'.[24]

There are very few lay-accounts of 'stabilization' in the 1920s and 1930s. This likely reflects the fact that until the closing decades of the twentieth century, the voices of people undergoing medical treatment were not usually considered particularly important. Few ever thought to listen to what they had to say, beyond, perhaps, general expressions of gratitude. The theory, practice, and history of medicine were the theory, practice, and history of great doctors, scientists, and researchers, and their patients were little more than a canvas. Even where we do find examples of individual people diagnosed during the 1920s and 1930s talking about their experience, many have clearly internalized this passive role. Their testimony tends to be quite matter-of-fact, and focuses on the technical training they received – being shown how to inject by practising on an orange, for example, comes up often.[25]

Hospital admission on diagnosis was, however, common practice for most of the twentieth century. There is a great deal more interview testimony from those taken in for 'stabilization' from the 1940s and beyond. An important caveat – we must be conscious that these accounts reflect the experience of those diagnosed some time after introduction of insulin. Nonetheless, their vivid descriptions of the hierarchical culture of the wards are often illuminating. 'Grace', for example, remembers being 'afraid' when she was admitted as a teenager in 1947:

> In those days, doctors . . . you didn't ask doctors. They were kind of a local god. You didn't argue or enquire. I don't think he would have put up with it. He was used to saying what would happen, not interested in what you thought. I never in those days [said] 'I don't want to do this' or 'why am I doing it?' You accepted [that] what the doctor said went . . . I was given a diet sheet to go home with [and] I was told that if I didn't learn to inject myself, I couldn't go home, and not [to] question anything.[26]

A full two decades later, Gillian Clifton's experience was similar. Like many diagnosed as children, she remembers being separated from her parents and committed to a ward for three weeks in 1967 at only six years old. Despite being visibly upset and asking to go home, she was initially told little about her new condition, its severity, or how long she would have to remain in the hospital. Once again, the doctors – almost all of them white, male, and upper middle-class – were unmistakably in charge:

> He wouldn't talk to you as the patient, but he would murmur things to the sister. There was no nurse, you know, it had to be the sister or the matron, and the younger doctors – everybody followed him around. It was like, you know, these disciples, and if he talked to you it was terrifying . . . They were very much captain of that ship and you didn't argue, you didn't disagree . . . They were godlike.[27]

The idea that doctors were all-powerful – to the point that they were considered almost akin to deities within their sphere of influence – comes up disconcertingly frequently. Carol Cowan was also diagnosed in 1967, and, like both 'Grace' and Clifton, reflected that they 'were gods . . . you just didn't challenge the medical profession'.[28] While these testimonies often also emphasize that the staff were kind, and clearly did care deeply about their patients, it is clear they expected to be obeyed unquestioningly.

This culture of authoritarianism was, perhaps surprisingly, even more palpable outside of the context of diabetes. Insulin also has a somewhat darker history as a psychiatric tool. First used to this effect by Austrian doctor Manfred Sakel during the late 1920s, by the 1950s insulin coma therapy (ICT) had become common practice across the globe.[29] This – as depicted in the 2001 film *A Beautiful Mind* – involved the deliberate administration of doses of the hormone large enough to induce coma, most often in those diagnosed with schizophrenia.[30]

The exact process differed between institutions, but it is possible to outline a typical routine. Insulin was injected in the morning, causing the patient to become comatose. They would then be maintained in this state for a variable period of time – often several hours – before being revived with glucose via injection or feeding tube. People subjected to this treatment, almost all of them institutionalized and powerless, underwent the procedure regularly, often with only one or two 'rest' days each week. This continued for as long as the attending psychiatrist decided it should – weeks, months, and perhaps even longer.[31]

Many mid-twentieth century practitioners felt that this technique was highly effective at treating the symptoms of mental illness. Nonetheless, the side-effects could be severe, and there are several examples of staff struggling to rouse their patients despite administering glucose.[32] Sometimes it caused permanent brain damage, and in rare cases people simply died.[33]

While it remained highly popular, by the mid-1950s the utility of ICT had become the subject of growing controversy. Harold Bourne's 'The Insulin Myth', for example, is one of the best remembered critiques of the period, appearing in *The Lancet* in 1953. Bourne did not oppose 'shock' therapies per se – in fact, he openly endorsed electroconvulsive treatment. Nonetheless, he argued that ICT was simply not effective, and that it was difficult to justify the considerable resources that it required – it was usually conducted on specialist wards with dedicated staff, and patients were monitored constantly much as they would be in an intensive care unit.

Controversially, Bourne even later suggested that one of the things that made ICT attractive to psychiatrists was that it gave them 'something to do . . . [making] them feel like real doctors instead of just institutional attendants'.[34] Deborah Blyth Doroshow, the historian he was speaking to when he made this claim, agreed with his interpretation. While they no doubt *did* believe that it worked, the clinical, involved nature of ICT also, she argues, served to lend legitimacy to psychiatry as a discipline, casting its practitioners as equal to more traditional doctors who specialized in wounds and physical diseases.

Despite these criticisms, ICT remained commonplace into the 1960s. When it was gradually phased out in favour of new antipsychotic medication during this decade, the transition was not always welcome. Even today, a few older psychiatrists look back fondly on the period, and believe that, in its time, it did a great deal of good. 'Of course there were dangers', claimed one as recently as 2014, 'but in those days the alternative was incarceration in a locked ward in a Victorian asylum, with little hope of rehabilitation or discharge.'[35]

Whether many of those who actually underwent ICT would be so sentimental is doubtful. Testimonies from former patients highlight quite how unpleasant the ordeal could be. Arthur Cain, who Doroshow interviewed in 2003, for example, recalled the experience of 'struggling to come back to life . . . from death's door in the coma'.[36] Don Weitz, who went on to become an outspoken critic of psychiatry as a discipline, recounts how, when he was institutionalized for fifteen months in the early 1950s, he begged his psychiatrist to stop 'torturing' him.[37] His protests fell on deaf ears.

Unsurprisingly, ICT, along with similar 'shock' treatments like electroconvulsive therapy, came to serve as a potent symbol of deep-seated problems within contemporary psychiatry to the discipline's critics.[38] For example, R.D. Laing, whose belief that most mental illness was social in nature and should be treated as such put him at odds with many of his colleagues, was directly influenced by his experience in the British Army Psychiatry Unit, where he spent some time working in the ICT division as a young intern.[39] He opposed the practice throughout his career, viewing it as yet another way in which his field worked to assert social and ideological control by aggressively pathologizing and 'treating' dissent.

Those who experienced ICT may have found the notion that people with diabetes considered their treatment with the same hormone unpleasantly authoritarian somewhat absurd, and given their marginalized position this is understandable. However, the divergent ways in which healthcare professionals asserted their power towards ICT patients and people with diabetes respectively speaks a great deal to the character and implications of insulin itself.

In the mid-twentieth century, psychiatric patients were often essentially prisoners. Once committed to an institution they were rarely able to leave before the staff decided to release them, and in the meantime they were forced to endure whatever was done to them. Their material reality was one of physical authoritarianism and control. Their bodies, while resident in the hospital, were not their own. To a lesser extent, this was also true of people with diabetes. While they did have more scope for discharging themselves, in order to receive the treatment that their life often depended upon they also needed to subordinate themselves to the authorities whilst on the ward. The big difference was that this group would, everyone accepted, return to their homes in relatively short order, and when they did so they would still have diabetes – they would continue to rely on insulin in their day-to-day lives, far from the watchful eyes of any professional.

In this context, the aggressive performance of authoritarian theatre served a pragmatic function. Knowing that they lacked practical power was precisely why physicians so consistently adopted an exaggeratedly moralistic, paternal tone with their diabetes patients. It was not enough to instil them with solely practical knowledge. It was also necessary to ensure, as best as possible, that they would remain *patients* by continuing to do what was expected of them. This is not to say that those with diabetes were expected to show no initiative. In the first edition of *The Diabetic Life*, Lawrence reiterates several times that at no point should the person using insulin change their prescribed diet (except to remedy hypoglycaemia), but he does suggest that, once sufficiently accomplished, they might 'learn to vary ... insulin according to the urine tests', but only extremely cautiously, according to his precise instructions, and only a few units at a time.[40]

'Peter', whose brother had been diagnosed in 1945 at the age of eight and had been treated by Lawrence, later described how he used to emphasize the importance of education and individual engagement with treatment:

Most of all, [my parents] learnt early on that a diabetic really cannot leave it to professional experts to control his diabetes for him. The diabetic has to know how to run his own life from hour to hour and day to day, and the role of the specialist is to provide the specialist knowledge, and the diagnostic tests that are necessary, in the context of the diabetic using his own head most of all . . . Lawrence was very clear indeed that the only hope for a diabetic was to be in charge of his own life.[41]

This was, however, understood in a highly qualified fashion. Small adjustments where necessary were one thing, but anything beyond this was strongly discouraged. 'Most of my patients have learned in time to make these adjustments most successfully themselves', Lawrence claimed in 1925, but it depended on 'the patient's intelligence and understanding of his disease whether this should be encouraged or forbidden'.[42] Nonetheless, throughout every edition of *The Diabetic Life* he continued to emphasize that people with diabetes 'must, however, not become too independent of [their] doctor's control'.[43]

Life on Insulin

But what exactly was prescribed? What was life on insulin actually like? While it may come as a surprise to those who have never required daily injections, the practical labour of diabetes management was rarely the source of significant stress in itself. With some exception, injections – even with the bulky, reusable glass-and-steel syringes used in the early twentieth century – quickly became tolerable for most.

Similarly, the daily process of assessing and recording urinary sugar content was simple enough. Urine was mixed with a reagent compound – usually Benedict's solution – and boiled. This would cause it to change colour. If no sugar was present it would turn blue, a trace amount would become green, and higher concentrations would be indicated by yellow or red. From 1945, Clinitest tablets simplified the procedure, and needed only to be dropped into a mixture of urine and water. While the risk of burning a hand on a hot test tube was always present, urinary testing itself posed little difficulty.[44] Some, children especially, even found the process entertaining.Clifton, for example, thoroughly enjoyed the chemistry of it all as a young girl, fondly remembering how 'it was like Jekyll and Hyde, because it all frothed and fizzed and rose up the tube'.[45] Nonetheless, 'poor' results could inspire feelings of guilt and shame. John Meredith, who was diagnosed in 1959, for example, even framed this in religious terms, reflecting that when the Clinitest solution turned 'orange . . . you have committed the sin. Full of sugar.'[46]

The source of such 'poor' results, of course, was often – in one way or another – food. Diet was by far the most onerous part of diabetes management for most diagnosed in the twentieth century. In order to prevent fluctuations in blood sugar, strictly regulated mealtimes and nutritional intake were considered essential. But what, exactly, did this mean? What were people expected to eat?

When insulin became available, physicians began to change their approach to dietary restriction only very cautiously. They often remained instinctively averse to allowing more nutritional intake than was absolutely required, and dietary allowances in the early 1920s continued, as a result, to be very low in carbohydrate. As always, individual physicians varied, but almost all agreed that, for the majority of their patients, most calorific intake should come from fats and proteins.[47] In some cases, they insisted upon less than one hundred grams of carbohydrate per day – far short of the 260 grams that Britain's National Health Service (NHS) now recommends an adult of average size and weight consume.[48]

This made sense according to contemporary understandings of the cause of diabetes, because additional food implied additional pressure on the islet cells. It also, however, possessed a palpable moral dimension. Prototypical New England Presbyterian Joslin, for example, felt that the self-denial and stoicism demanded by dietary control meant even more when, theoretically, insulin could patch over the consequences of immoderation. By continuing to strictly abide by direction, his patients demonstrated their humility and, implicitly, their worthiness of continued life.[49]

Equally important, however, was simple economy. In some locations, political intervention made insulin broadly accessible. In 1924, for example, *The Lancet* wrote approvingly of the Ontario provincial government's decision to provide people with diabetes 'free insulin on the certificate of their medical advisers'.[50] Such examples, however, were the exception, not the rule. Most people with diabetes were required to purchase what they needed.

When Eli Lilly began to sell 'Iletin' wholesale in early 1923, as per their agreement with the University of Toronto, it went for around $5 per hundred units – around $80 today. In Britain, the same amount cost around twenty-five shillings from April 1923 – the equivalent of about £50 adjusted for inflation. Given a typical prescribed diet at the time might have called for up to twenty or so units daily, and sometimes more, this was an expensive medication indeed.[51]

Robert Tattersall – who, as we shall see in the following chapters was an important figure in his own right – wrote extensively on the history of diabetes after his retirement from medicine. In one article, he highlighted the exorbitant costs involved in these early years by pointing out that someone living in London in the spring of 1923 might have bought six large bottles of imported Portuguese wine for less than a week's supply of insulin.[52]

Unsurprisingly, many voiced concerns that wealthier people with supposedly 'mild' cases might selfishly purchase vast amounts of insulin to eliminate their symptoms and enable them to abandon any dietary restriction despite being in little immediate danger, and in doing so deny it to those on the brink of death. Britain's Medical Research Council (MRC) was so worried about this, in fact, that it insisted early in 1923 that there would be no 'luxury use'. Those with the most 'severe' symptoms would be prioritized where possible, but even they would get no more than they absolutely required. The announcement was well received. One doctor wrote that:

The supply is not yet large, and I note with great satisfaction a regulation laid down by the Medical Research Council. There are many persons who are quite well so long as they avoid starch and sugar, but if they consume foods containing those things the symptoms of diabetes will appear.

Many such persons are well-to-do – this is a disease principally of the well-to-do – and they would gladly pay half a crown a dose for insulin for the satisfaction of eating what they please. Meanwhile, persons lying in the shadow of death from diabetic coma might be lost for lack of the dose which a wealthy gourmand had appropriated. Until there is insulin for all, the M.R.C. have forbidden its 'luxury use', and we must applaud that humane prohibition.[53]

Insulin prices did begin to decline by the end of 1923, but it remained a significant financial burden. By the summer of 1924, Connaught was charging a dollar per hundred units, while Britain's MRC had reduced the price in the UK to two shillings for the same amount the previous March – the equivalent of about C$15 and £5 respectively today. This was an impressive reduction, but costs could still soak up a substantial portion of a working person's pay packet. An unskilled male labourer in England at the time, for example, might have brought home only around forty shillings per week.[54]

In addition to theoretically preserving the islet cells, limiting carbohydrate meant that the amount of insulin needed by each individual could be minimized. This was one way to provide treatment while also doing what was possible to control the financial burden that it implied. Nonetheless, costs continued to be an impenetrable barrier for some. In a 1927 article in the *BMJ*, for example, one doctor admitted that 'patients who were insured persons were, of course, able to obtain supplies free, but in other cases the supply had often to be limited, or its use entirely discontinued on the grounds of expense'.[55]

Even for those who could easily afford their medication, prescribed diets were usually meagre. The experience of Alan Nabarro provides one good example of this. When he became one of Britain's first insulin recipients in April 1923, Nabarro was fortunate to have been born into a wealthy middle-class family, and he never had reason to ration his supplies on economic grounds. Nonetheless, his diet sheets from the mid-1920s show that he was still expected to follow a very strict and highly restrictive meal schedule. He was prescribed only 80 g of carbohydrates per day, approximately the same amount of protein, and an incredible 140 g of fat.[56] Despite the limitations placed on him, Nabarro's experience was eased by his more intangible privilege. As a precocious, well-educated child with several prominent doctors in his immediate family, he was able to develop a long-running, mutually respectful, and even occasionally humorous correspondence with his diabetologist Geoffrey Harrison – the same Harrison whose telegram had saved Lawrence's life. The letters between the two show a clear mutual fondness, with the older man occasionally jokingly signing off 'your old enemy, doctor'.[57]

Harrison was wryly alluding to the fact that his young patient was not always enthusiastic about the demands of his prescribed treatment, but the familiarity it represented was useful. Before Nabarro attended his Bar Mitzvah, he was able to ask for dispensation to enjoy a (very slightly) more luxurious meal for the occasion, and he requested advice as to how best to administer his insulin to compensate for this. Harrison happily provided this, and did similar when he later sought guidance for eating out and attending the theatre.

Despite his privilege, the few indulgences Nabarro were permitted did not amount to much. His Bar Mitzvah diet sheet, for example, differs only marginally from his usual prescription.[58] The boy – barely in his teens – was consuming around a third of the daily carbohydrates now recommended for an adult, but almost twice as much fat. Worse, he ate almost exactly the same thing, at the same time, every day, with the only significant variable the meat component of each meal. On the whole, his food was repetitive and bland.

One of the reasons for this banality was probably practical necessity. In the 1920s, packaged food was significantly less common than it is today, and what did exist did not generally come labelled with the kind of precise nutritional information to which we are now accustomed. Bread, milk, and other simple foods could easily be weighed and measured without any complex calculations, and a given measure could be relied upon to provide a definite amount carbohydrate, for example, or protein, or fat. This was not easily achieved with more complicated meals, which might involve a range of different ingredients. Some, Nabarro among them, kept a few recipes for more involved foods with precise cooking directions – his documents include instructions for making marmalade pudding and apple charlotte, for example – but these were clearly occasional treats.

From the mid-1920s, this could be somewhat mitigated by the use of dietary substitution plans. One of the earliest and best known of these was the 'line-ration' system devised by Lawrence, who in characteristic fashion did take issue with boring diets that left people dissatisfied and hungry. He believed that any 'satisfactory diabetic diet' should be constructed around four criteria, and expressed frustration that most of those he saw prescribed fulfilled only the first two:

1. It must supply the correct calorific requirements of the individual, being neither too low nor too high.
2. It must contain the proper proportions of carbohydrate, protein and fat.
3. It must satisfy the patient in quantity and quality as far as possible.
4. It must be accurate, simple to calculate, and permit of variety.[59]

The 'line-ration' method was created to meet each of these demands, and provided an elegant solution to dietary repetitiveness. According to this system, prescribed food was not expressed in exact figures, but rather a certain number of black and/or red 'lines'. Black lines represented a fixed portion of carbohydrate, red ones protein and fat. Instead of being told to have, say, twenty grams of carbohydrate, fifteen of protein, and eighteen of fat, a mealtime allowance might simply say 'two black lines and two red lines'. What made this particularly liberating was that *The Diabetic Life* – later supplemented by resources produced by the BDA – included a list of different measured food items that each made up a single red or black line, and meals could utilize any combination of these. An egg, for example, could be exchanged for ¾ oz of cheese, or 1 oz of lean bacon, by referring quickly to the easy reference table provided.

Nonetheless, eating continued to be the source of much of the stress that came with living with diabetes. As the point at which the condition most closely intersected with everyday life, the need for special treatment could emphasize feelings of otherness. 'Ann', for example, was diagnosed with diabetes in 1946 as a child, and remembers being acutely embarrassed at having to take a pair of scales into school with her to weigh her lunch. Her fellow pupils, especially the younger ones, 'all wondered what on earth [she] was doing, why [she] had to do this', and she found it very difficult to explain in a way that they would understand.[60]

Even in the more understanding environment of a caring home, the mark of difference was often unavoidable. Clifton, for example, remembers her mother sterilizing her needle and syringe at breakfast.

Whether to save time, as a genuinely kind-hearted act of symbolic acceptance, or perhaps both, she used to place the equipment in the same pan that she was using to boil the family's eggs.[61] In doing so, however, she implicitly highlighted that Clifton was not quite the same as her siblings.

That people on prescribed diets had to force down their meals to match their insulin could also have a significant psychological impact – particularly where both high and low blood sugar were moralized as personal failure. This continued to be an issue well into the late twentieth century. Lis Warren, a well-known British diabetes activist, remembers how she used to resent eating her cereal every morning after her diagnosis as a teenage girl in 1965:

> If you were told your daily dose was eighteen units that's what you had. So you got up in the morning and you had to then eat to feed that dose. So if you got up and you weren't hungry you had to force-feed yourself, and that affected me a lot. So, if you have a normal cereal bowl, that might have been two or three portions of carbs, but I was on five portions for breakfast, so I had to weigh my Rice Krispies and I used to serve it in the kind of serving bowl that you'd make a trifle in, or something for a family! A big bowl! Every day I used to sit down to a big bowl of cereal like that to get my fifty grams of carbs and I wasn't hungry! I didn't want it![62]

Forcing down large portions of food to meet requirements could also have physical consequences, which often acted to amplify the psychological ones. Joanne Pinfield, who was diagnosed in the late 1970s, for example, remembers how having 'the same amount of carbohydrate every meal' could often be 'very hard if [she] wasn't hungry'. Because she forced herself to eat more than she was comfortable with, Pinfield put on weight. Cruelly, she was ridiculed for it by her classmates. While the bullying made her self-conscious, it was in reality the demands of her treatment that contributed most to her sense of powerlessness – 'I had absolutely no control over what I had to eat, so that was a difficult thing to cope with.'[63]

Unsurprisingly, people with diabetes often developed disordered relationships with food. Clifton, for example, half-jokingly claims that she still cannot 'look a Weetabix in the face' after years of being fed nothing else for breakfast. As a teenager her dietary restriction came to have serious consequences. For her, food became invested with a double meaning. It was, paradoxically, both cure and cause, and she regularly ended up in a spiral of binge-eating and guilt:

> I wanted to eat but it felt like it was wrong to eat. It felt like it was wrong to eat because I had diabetes. It was wrong to eat because it would make me fat. Food became a bad thing, and certainly, at the height of me having these issues with food, I could think 'right, I'm not eating today', and I would monitor my blood sugar and everything. But if I just had an apple, that was it. I've blown it so now I might as well eat what I want . . . It felt like it was almost sinful and something I had to do secretly.[64]

The mental health impact of diabetes and its management was rarely addressed in early to mid-twentieth century professional writing. Lawrence brushed the topic off in a few lines in 1965, claiming that despite 'the somewhat abnormal, or at least unusual, nature of the diabetic life, mental balance is surprisingly little upset or the character changed, or strain or depression evinced'.[65] Given the above testimonies, however, this is almost certainly a far cry from the truth, and likely reflects the boundaries of what was, and what was not, spoken about in the context of the clinical relationship during his days in practice. Today, it is well recognized that diabetes can play a major role in contributing towards various mental health conditions and eating disorders.[66]

This is perhaps best demonstrated by a condition known as 'diabulimia', sometimes described as 'the world's most dangerous eating disorder'.[67] Most common in – though by no means exclusive to – young women, this, in short, involves people with T1DM deliberately using less insulin than they need. By doing so, they attempt to harness the emaciation that comes with lack of treatment to lose weight.[68] 'Diabulimia', unsurprisingly, is extremely dangerous. Those affected do themselves great physical harm, and the psychological distress that can come with it is considerable.[69]

Despite his nonchalant attitude towards mental health, Lawrence was optimistic about the long-term prospects of people with diabetes.

Until 1941 he even felt comfortable claiming that 'the faithful diabetic is sure of his reward in health'.[70] Throughout the 1940s and 1950s, however, it became increasingly obvious that people on insulin were, much like their often older, diet-treated equivalents, vulnerable to dangerous long-term complications like blindness and kidney disease – a subject that Lawrence began dedicating a whole section of *The Diabetic Life* to from 1955.[71]

Like his colleagues, however, Lawrence was at a loss as to why, exactly, these long-term problems occurred. In practice, he had always felt that keeping blood sugar within a 'normal' range was important, largely due to his lifelong belief that hyperglycaemia 'overstrains' the islet cells. Perhaps unsurprisingly, this ethos also influenced his response to the new challenge. In his experience, it seemed that 'the wildest and most uncontrolled cases' were most keenly affected.[72] Implicitly, therefore, careful management was the best insurance.

Lawrence believed that 'on the whole, *prognosis* depends on *continuous correct treatment*', facilitated both by comprehensive self-education and committed professional supervision.[73] But did any of this actually matter? These long-term issues, even he admitted, seemed 'mysterious' in the mid-twentieth century, and no one could say 'why an excess of circulating sugar, if it does so, should induce [them]'.[74] Could it be that many physicians were simply being overzealous in their efforts?

Free Diets

A decade after the introduction of insulin some physicians had begun to modestly liberalize their treatment strategies. In 1934, Harold Himsworth wrote that 'on its first introduction insulin was given merely as an adjunct to the Allen diet, but as its use became more widespread it began to be employed in conjunction with diets containing larger and larger quantities of carbohydrate'. Doing so had been a controversial process, as it appeared to completely contradict the received wisdom that such an increase in volume would overwork 'the diabetic's already diseased islets ... [causing] rapid deterioration of the clinical condition'. In practice, though, this did not seem to happen. The majority of those on higher carbohydrate diets reported feeling 'much improved', even after some time, and in most cases no additional insulin was required for a relatively small increase in allowance.[75] This seemed puzzling, and Himsworth concluded by suggesting that perhaps the issue lay with insulin sensitivity. While the exact mechanism at work was, he admitted, not clear, it seemed undeniable that 'a carbohydrate diet facilitates the development [of higher sensitivity]; a fat diet retards it.'[76] By consuming proportionally more carbohydrate and less fat, a given dose could actually be made more efficient, allowing for far more satisfying meals at little to no clinical cost.[77]

By the late 1930s the unenviable, fat-heavy diets of the early 1920s had usually been moderated considerably in practice. Economic factors certainly had a hand in making this possible. Insulin costs had fallen dramatically. In 1942, Connaught, for example, was selling its product for only C$0.20 per hundred units – around C$3.35 today. This had made it considerably more accessible to those from poorer backgrounds. In 1939, for example, working-class schoolboy Charles Lewis had a prescription of 20 units of soluble insulin twice per day, alongside 130 g of carbohydrates, 65 g of protein, and 105 g of fat – still undeniably restrictive, but 50 g more carbohydrate and 35 g less fat than the far wealthier Nabarro had been permitted some fifteen years earlier.[78]

Another important development was the introduction, from the mid-1930s, of several modified versions of insulin that acted on the body for longer than the soluble type. Hans Christian Hagedorn, the scientific lead at Denmark's Nordisk Insulin Laboratory, produced the first in 1936 using protamine – an extract of fish semen – and by the mid-1950s several varieties were available, for example Protamine Zinc Insulin (PZI), Neutral Protamine Hagedorn (NPH), and Lente.[79]

These new insulins were received enthusiastically by many physicians. In theory, it had now become possible to manage diabetes using only a single daily injection. Early experiments, however, seemed to show that this was often more trouble than it was worth. In 1939, Lawrence admitted that 'with this treatment, in severe cases, it is rarely possible to keep the urine *constantly sugar-free without risk of hypoglycaemia*'.[80]

But how much did any of this matter? Why bother to control blood sugar levels at all? The vast majority of physicians in the early insulin period believed that keeping urine tests free of sugar should be the primary goal of therapy, but there was little clear evidence that such a strict approach had any real influence on long-term health. By the 1940s, some were pushing back against the idea that high blood sugar was harmful at all.

Chief among these critics was the New York Hospital's Edward Tolstoi. In the late 1930s, Tolstoi had begun to incorporate PZI into his practice, hoping to develop an approach to treatment that required only a single dose of insulin. This, he felt, would permit his patients 'more freedom as it would obviate multiple injections and free him from the slavery of the definite time relationship between the insulin administration and his meals'.[81] However, like others, he found it difficult to use effectively. It often led to sugar in the urine, but increasing the dosage and incorporating additional shots of soluble insulin resulted in frequent episodes of hypoglycaemia no matter how much he juggled with the prescribed diet. His efforts had achieved precisely the opposite of what he had intended – 'the patients were receiving multiple injections of insulin, they were burdened with additional dietary instructions and they lived in apprehension as [hypoglycaemic] reactions were unpredictable if the urine was kept sugar free'.[82]

When he had begun to experiment with PZI, Tolstoi had asked his patients to check in to his clinic weekly so that their progress could be monitored. Several, however, had not. A handful returned over the following weeks to explain that they had 'never felt better and stronger', and that they had only come back at all because their urine tests showed consistently high levels of sugar – something most people with diabetes were trained from the outset to recognize as a bad sign.

Tolstoi was intrigued, and had two of them admitted to the hospital's metabolism ward where they could be observed more closely. At a conference in 1940, he reiterated his surprise at the results. Despite urine loaded with sugar 'they were amazingly free from any and all symptoms of diabetes mellitus, they maintained their weight . . . and the urine was free from acetone and diacetic acid'.[83] They seemed, essentially, to be in perfectly good overall health. Tolstoi later tried to replicate the results, selecting 'one of the most severe and practically uncontrollable diabetics, if a normal blood sugar and freedom from sugar in the urine are considered the ideal' for the trial.[84] Once again, he felt the outcome positive:

A severe case of diabetes mellitus is presented in which the treatment with multiple injections of insulin, aiming at a normal blood sugar and sugar free urine, failed to produce as good clinical results as a single dose of protamine zinc insulin, with no attempt to abolish hyperglycaemia and glycosuria ... Under constant experimental conditions, the glycosuria showed enormous and unpredictable variations. Despite the heavy and continuous glycosuria for three years the patient has not developed any more colds or other infections, than the non-diabetic. His renal function shows no impairment, his atherosclerosis no demonstrable increase.[85]

By 1943 Tolstoi was entirely convinced that high levels of blood sugar were completely harmless. His experience led him to turn his approach to practice upside down. Throughout the 1940s, he began to advocate for the use of a 'free diet', arguing that if his patients did not seem to experience any real privation despite the presence of sugar in the urine, there should be no reason to avoid it. Strict prescription, he now suggested, was a relic of the pre-insulin era. Contrary to popular belief, he argued, people with diabetes should be able to eat essentially whatever they liked whenever they liked, taking a single shot of PZI before 'putting their syringes away for the remainder of the time as they would their toothbrushes'.[86]

This approach was met with fierce criticism from more conservative figures. Joslin, for example, argued in 1940 that high blood sugar was a 'fundamentally abnormal state', and that ignoring it was akin to failing 'to heed the red signal at the railroad crossing'.[87] Tolstoi, however, dismissed the attack. A decade later, he continued to argue that most of Joslin's patients, too, developed complications eventually despite the standard that he held them to.[88] This, he thought, was just one of the inherent perils of life with diabetes.

'Free diets' were undoubtedly an attractive proposition to those who had previously had to prescribe, not to mention those who had to live with, strict dietary regulation, and Tolstoi enjoyed several years in the limelight. He published several articles on his approach to treatment in mainstream journals, and contributed a chapter to the edited volume *Progress in Clinical Endocrinology*.[89] His success was, however, was to be short lived. By the mid-1950s, new research seemed to show that 'free diets' could, in fact, be extremely harmful indeed.

One of these studies, surprisingly, was conducted in the UK, where Tolstoi's ideas had never really caught on. In the late 1940s, the Edinburgh Royal Infirmary played host to a five-year trial of the 'free diet' concept. When the results were published in 1951, the outcomes seemed relatively positive.[90] In 1954, however, Derrick Dunlop, one of the original authors, provided a grim update. The experiment, he now felt, had in hindsight been 'disastrous' for those who had taken part. Of the fifty initial participants, only nine remained in good health a decade after adopting a 'free diet'.[91] Dunlop explained how the experience reinforced a more traditional attitude in his practice:

> As the result of this and my experience of 'free diets' I have returned to my original simple diabetic faith. I believe that whatever specific aetiological factors [underlying causes] may be causing diabetic degenerative lesions ... the careful control and aggressive treatment of the disorder over the years is a most important factor in their prevention or postponement. I believe that to obtain good control diabetic diets should not usually contain much more than 200 g of carbohydrate; that patients should be initially trained in the hard school of food-weighing ... and that they should report regularly to a diabetic clinic to be assessed ... and, depending on the findings, to have their insulin dosage and diet suitably altered, for it is most exceptional to encounter a well-controlled diabetic who has been made entirely responsible for his own treatment.[92]

While he and a few other holdouts continued to protest, Tolstoi soon became isolated within the profession. Nonetheless, despite the now-obvious weaknesses of his approach, we should not dismiss him wholesale. His work contained a valid and important implicit critique of paternalism in medicine. Tolstoi consistently took issue with the strict authoritarianism of figures like Joslin, feeling that their approach was pointless and, worse, counterproductive, only encouraging mutual distrust and resentment. In their personal lives, he claimed, few people actually kept to their prescribed treatment for long. Instead, they deviated as their circumstances demanded. In many cases, the reality of everyday life made it difficult to remain faithful to instruction – particularly where fear of discrimination meant many would do whatever they could to avoid being 'known as a diabetic'.[93] As we shall see in the following chapters, this assessment was often correct.

By failing to understand the complex social context within which insulin therapy was conducted, condemning any deviation from diet whatever the reason, and chastising patients accordingly, Tolstoi believed that frustration emerged on both sides. As a result, 'a system of deception developed with the patient and the doctor', which made effective treatment almost impossible:

> When the patient visited the clinic we were satisfied if we found that he was well 'controlled' and we were naive enough to assume that he was also well 'controlled' between visits. We learned quite frequently that a patient would leave the clinic, after having been complimented on his excellent cooperation, and would go at once to the hospital cafeteria for coffee and doughnuts or chocolate cake and sometimes we found him enjoying an ice cream soda. We then looked the other way while saying to ourselves, 'Oh well, he was sugar free.'[94]

Tolstoi tried to avoid this pitfall by understanding those under his care not as automatons, but as complex human actors upon which medical demands were only one pressure. Even before he had fully devised the concept of the 'free diet', his practice seemed astonishingly 'patient-centred' for the early 1940s. He was, for example, always wary of admission for 'stabilization' where it was avoidable, and he always included his patients in the decision-making process when it came to designing dietary regimens.

However, the discrediting of the 'free diet' movement implicitly discredited everything else that Tolstoi had done differently. Reflecting on the Edinburgh trial, for example, Dunlop was also making an ideological point. While acknowledging that persistent high blood glucose could be harmful, he also reasserted the paternalistic authority that he felt should characterize the clinical setting – an authority that figures like Tolstoi had threatened to upend.

The Reassertion of 'Orthodoxy'

By the 1960s, the fallout over 'free diets' had led most practitioners to go the way of Dunlop, and to return to an ethos of diabetes management similar to that advocated in the 1920s and 1930s. Few now challenged the importance of maintaining stable blood sugar levels, or disputed that strictly following a prescribed diet was the optimal way of achieving this. The way diabetes was typically approached at the time in Britain, for example, is effectively portrayed in *Living with Diabetes*, a short film produced by the Wellcome Foundation in 1959.

The film follows two newly diagnosed characters as they are taught how to manage their condition, both of whom serve as caricatured stereotypes of people living with diabetes. The thin, modest, and deferential 'Miss Smith', for example, is contrasted by the loud, overweight 'Mr Anderson'.[95] They are guided by Geoffrey Lewis and Stella Riley, who together encapsulate the highly gendered character of medical practice at the time of production – he an immaculately dressed, middle-aged consultant with a slightly patronizing, fatherly demeanour, and she a stern, matronly dietician.

Throughout *Living with Diabetes*, both characters are frequently reminded of the importance of following their prescribed treatment to the letter. The importance of this is made even more explicit towards the film's conclusion. In one sequence, 'Miss Smith' decides to forgo her evening meal to go dancing. Consequently, at the close of a comical scene in which she spins disoriented around a dancefloor, she realizes that she is experiencing an episode of hypoglycaemia, sending her bemused partner to the bar to fetch sugar-laced tea before drinking it, deflated, as partygoers continue to enjoy their night around her.

'I don't think I'll make that mistake again', she later reports to Lewis with the air of a remorseful child. 'No, I don't think you will either', he responds with unmistakable condescension, concluding the segment with an earnest reminder that 'it may seem fairly easy for diabetics to make a reasonably good job of looking after themselves. And so it is, provided you stick to the rules!'

In 1962, the authors of *The Story of Insulin: Forty Years of Success Against Diabetes* agreed with Lewis' assessment. They declared that:

The diabetic death has been replaced by the diabetic life ... Insulin does not cure, it alleviates – but to such a degree that the patient can lead a normal existence and live a good measure of years. All this is possible provided he does not neglect the simple rules of daily living that are laid down for him.[96]

There is a cautiously optimistic tone to this book, but its accuracy seemed increasingly doubtful. Insulin saved people from imminent death, but it was becoming clear that they rarely lived close to as long as the general population no matter how carefully they followed their doctor's orders.

As early as 1950, Ruth Reuting, a colleague of Joslin in Boston, had published a foreboding analysis of the long-term prognosis of insulin recipients. Of fifty young patients at the city's New Deaconess Hospital followed since 1929, nineteen had died at an average age of just under 35, with cardiovascular and/or renal failure the leading cause. Of the surviving number, only four seemed in generally good health.[97]

Over the following decades, more studies seemed to corroborate these depressing findings.[98] While the introduction of insulin had seemed a dramatic triumph, the latter half of the twentieth century was defined by marked pessimism. In 1930, a jubilant Allen had declared that diabetes had 'been scientifically mastered ... [and that] every patient can be expected to live out his full natural lifetime'. Looking back, however, this now seemed hopelessly naïve.[99]

Chapter 3

'Intensification', 1976–1993

While absolutely conclusive proof remained elusive, by the mid-1970s it had become very difficult to dispute a connection between overall blood sugar levels and the long-term complications associated with diabetes.[1] In 1976, the American Diabetes Association (ADA) altered its policy to reflect this. It now seemed clear, Harvard Medical School's George Cahill claimed, that 'the goals of appropriate therapy should thus include a serious effort to achieve levels of blood glucose as close to those in the nondiabetic state as feasible'.[2] This, however, was much more easily said than done. In a 1949 lecture, R.D. Lawrence had argued that 'anyone who confidently claims to maintain a physiologically normal blood sugar in the average insulin case has, I am sure, no wide or accurate experience'.[3] By the late 1970s, this remained the case, with one textbook arguing that 'all control is poor in comparison with physiological levels, often very poor'.[4]

As things stood, the situation looked grim. Implicitly, everyone using insulin was at serious risk whatever they did. Cahill understood this, arguing that 'current means of therapy are only partly effective at best, and therefore a high priority must be assigned to the development of more physiologic insulin delivery systems or to approaches to the correction of the deficient insulin-producing mechanism itself'.[5]

Mainstream contemporary approaches to management seemed to make this very difficult. People using insulin tended to be prescribed two, or perhaps at most three, fixed doses per day, alongside a strict schedule of meals. This approach, however, did little to replicate the pattern of insulin production seen in those with functional pancreases, and many of those who used it experienced frequent periods of very high blood sugar. If, as the evidence seemed to suggest, this made them more vulnerable to long-term complications, then a new strategy was necessary. Treatment needed to be adapted – or, in the professional turn of phrase, 'intensified' – to reduce the risk. How, exactly, this was to be achieved, however, was less clear.

Deus ex machina?

This seemed a very difficult position for many medical professionals. One of the major challenges was that insulin therapy was almost always conducted away from their supervision. It was as a result virtually impossible to accurately assess overall blood glucose levels on a day-to-day basis, much less come up with ways in which they might be brought down. While people were generally encouraged to perform urine tests at home and record the results, the information they provided was of limited value because sugar spills into the urine only once it passes a certain level in the blood.[6] There is no way of knowing exactly how high blood sugar is at a given moment by using urine tests, only whether or not it has risen beyond a certain level at some point in the preceding few hours.[7]

Direct blood tests were far more accurate, but these required laboratory analysis and were, as a result, taken only during clinical appointments in the first decades of insulin therapy. In 1965, however, Indiana's Miles Laboratories started to manufacture Dextrostix. These were thin paper reagent strips which, on contact with a small droplet of blood, would change colour to reflect its sugar content. The results could be difficult to interpret, so five years later the company released the Ames Reflectance Meter to do the job automatically.

Costing around $500 apiece – equivalent to $3500 today – this was an expensive, bulky machine, but it could read Dextrostix far more precisely than the human eye and gave results only marginally less accurate than a traditional laboratory analysis. With a reflectance meter, blood sugar readings could be acquired in a matter of minutes. In theory, it had become possible to perform regular, effective testing throughout the day, giving a much more useful picture of overall trends according to which persistent areas of concern could be identified.[8]

In practice, however, this did not happen for some time. In the 1970s, blood sugar measurement, unlike urine testing, remained widely understood as a job for the doctor or nurse. The Ames device was initially sold only to qualified medical professionals, and it was never intended for home use. A very small minority, such as engineer Richard Bernstein, who had his psychiatrist wife sign off on the purchase, were able to acquire one, but the machine was designed for a clinical setting.[9]

In this context, a decision made in 1975 by Clara Lowy of London's St Thomas' Hospital must have seemed outrageous. By the 1970s it was well-established that pregnancy could be particularly dangerous for people with diabetes, and physicians tended to emphasize the importance of maintaining stable blood sugar levels throughout to minimize the risk to both mother and baby. One of Lowy's patients, just entering her third trimester, had been admitted to hospital after a series of unexplained hypoglycaemic episodes. As was standard practice at St Thomas' in such situations, the plan was to keep her under strict observation on the ward until she gave birth.

Understandably, the woman was less than thrilled about the prospect of spending months committed to an institution. She asked why she couldn't simply do the necessary tests at home using a reflectance meter and bring the results to her usual outpatient appointments, where any necessary adjustments to treatment could be discussed. Lowy, in a move that left many of her colleagues shocked, acquiesced. The woman quickly learned how to use the meter effectively, had an otherwise uneventful pregnancy, and gave birth to a healthy child.

Afterwards, self blood glucose monitoring (SBGM) became standard practice for expectant mothers in Lowy's clinic, and many of those who started performing it found that it dramatically improved their quality of life while allowing them to maintain much more stable blood sugar levels. Despite this, some of her contemporaries thought the policy absurd, and even potentially dangerous. Undeterred, she stuck to her guns. This was a more significant development than it might first appear. By giving active responsibility for a specific diagnostic process to her pregnant patients, Lowy began to implicitly acknowledge the shortcomings of the paternalistic model of care.

Had she operated according to that traditional framework she would have been unable to achieve the positive improvements to clinical outcomes that she did. Diabetes management, this suggested, benefited from active collaboration with the patient, who, Lowy claimed, often became 'so expert' at self-treatment that 'they [were] able to alter their insulin dose' according to SBGM results.[10] Looking back in 1998, she admitted that the experience had highlighted – in a way that had seemed ground-breaking at the time – that 'patients may contribute more to advances in medicine than is recognised'.[11]

This posed a question: what if everyone using insulin was encouraged to test not only their urine, but their blood as well? Would this not provide an obvious way to gain more insight into day-to-day blood sugar trends, and perhaps permit them to be lowered? Inspired by Lowy, a number of other physicians began to trial the concept with their own patients. In almost all cases, it seemed that, contrary to popular belief, almost anyone could safely learn to test their own blood sugar at home, and provide records with perfectly adequate accuracy.

The medical establishment was not easily persuaded, however. Latter-day diabetes historian Robert Tattersall, for example, was one early advocate of SBGM. He submitted a paper on the topic to the BDA in advance of a meeting in 1977, but it was rejected. When he brought up the subject during the discussion anyway, he was almost laughed out of the room for his apparently ludicrous suggestion.[12]

Nonetheless, only a year later the dam broke. A series of enthusiastic articles on SBGM – Tattersall's included – were published in 1978, and more quickly followed.[13] Years later, in 2017, he – with no small hint of satisfaction – wrote that despite the initially lukewarm response, 'within five years it became standard practice'.[14] If anything, he was being modest. In January 1980 *Diabetes Care*, a journal published by the ADA, released a special edition on SBGM based on a conference that had been held in New York the previous year. In his opening statement, the chair seemed to echo the immediate response to insulin itself when he announced that 'it is not often given to us, as physicians, to witness consciously the beginning of a new era in medicine'.[15]

When it came to assessing long-term blood glucose levels, however, SBGM had one major flaw: it relied upon a human operator. Like urine tests before it, results could be falsified, and in the morally charged context of insulin therapy, they often were. Colin Dexter, for example, was diagnosed with T2DM in 1987, and openly admits that he made up his results on some occasions:

> I was never quite so honest or honourable about keeping a blood sugar [record] as I should have been. I used to do it two or three days before in a beautifully neat tabulated form and then extrapolate backwards, and never tell them if I was way high – I'd never tell them if I was high at all, or if I was low at all, really. I used to stick it in the middle . . . because I didn't want to upset my dear friends who were trying to help me all the time.[16]

Dexter frames his actions as part of an effort to protect his doctors from the 'upset' of 'unsatisfactory' results, but in some cases falsification served a much more immediately pragmatic function. Gillian Clifton, for example, remembers filling out her logbook with invented numbers the night before her clinic appointments as a child. To avoid detection, she avoided 'perfect' figures and included just enough high and low results to ensure that the record appeared plausible. Unlike Dexter, Clifton did this because she understood that the 'disappointment' of her doctor would be interpreted as a personal failure on her part, and likely lead to an angry lecture.[17] She even nicknamed one of her consultants 'Dr Dragon' for his propensity to fly off the handle when displeased. In this context, providing 'acceptable' results shielded her from being reprimanded for something that she felt she had little control over, whatever she did.

That people with diabetes sometimes falsified their results was not lost on medical professionals.[18] Concern about the implications of this was, in fact, so pervasive that a considerable amount of time and effort was put into research – some of it more than a little ethically dubious – that questioned why this phenomenon occurred and, more importantly, attempted to find ways to prevent it. Throughout the 1980s, for example, several studies were conducted in which participants were provided with what they thought was a standard blood sugar monitor.

In reality, the device had been modified to store test results without the user's knowledge, so that the electronic record could later be compared to the paper one.[19]

Such work did, at least, not go without justified criticism. In 1991, physician and sociologist David Armstrong explicitly asked what, precisely, this kind of research hoped to achieve. By expending so much effort trying to determine whether or not people could be 'trusted', he argued, these studies only helped to cultivate an adversarial relationship between doctors and their patients.[20]

The belief that a significant number of people were being deceptive with their SBGM records, however, meant that another technology – the HbA1c test – was welcomed by many professionals. This had first been used in 1976, when Ronald Koenig pointed out that excess sugar in the blood permanently binds to red blood cells. The greater the excess, the more this occurs, and, because the cells can live for up to four months, measuring its extent can provide useful long-term data.[21] In short, the HbA1c test shows average blood sugar values over the preceding three to four months. By the mid-1980s, it had become a common component of the array of tests performed during regular clinical reviews.

While it did provide useful data, the HbA1c test left people with diabetes with nowhere to hide. Whatever a logbook or isolated blood sugar reading said, physicians now had the long-term data necessary to corroborate (or not) this less reliable evidence. This, as Christiane Sinding argued, allowed them to effectively reassert moral control, 'embodied in a technical molecular device rather than being espoused in a "catechism" as in the early 1920s'.[22]

It is little wonder that Clifton, remembering the introduction of the technology, half-jokingly refers to the 'horrible, horrible person who invented it' – it was yet another source of judgement and stress.[23] British diabetologist Charles Fox recalls how the chair of a local BDA branch – someone living with the condition, not a doctor – reacted:

> He said to me, 'Oh, I'm not sure about this HbA1c thing, Charles . . . It's a spy in a cab. It's the spy in the cab!' So, the patients immediately knew what it meant, and most of them weren't pleased about it because they felt that confessional thing, where you went in and named your sin

and were blessed, and you went on. That rather comfortable relationship could be changed by somebody discovering that this perfect person with diabetes who always had negative glucose results . . . actually [had an HbA1c of] 12%.[24]

The possibility of immediate blood sugar measurement alongside access to data on long-term trends did, however, enable a dramatic expansion of research into different strategies for delivering insulin. Recognizing the importance of developing a more 'physiological' approach to management, and despairing at the limitations of contemporary treatment, physicians began to look elsewhere for something – anything – that might provide a solution. Perhaps unsurprisingly, many of them invested their hopes squarely in technological innovation.

One proposed suggestion was to jettison the strict timetable of daily injections entirely, and instead use a pump to infuse insulin directly into the body as required. In fact, this was not a new idea. In the early 1960s, Arnold Kadish, a doctor in Los Angeles, had published a series of papers on what he described as a 'servomechanism for glucose monitoring and control'. The device – the size (and weight) of a backpack – was directly connected to the user. When in operation, it automatically kept track of their blood sugar levels and, when they became elevated, activated the attached pumps to deliver onboard insulin. The machine was impressive, and Kadish certainly felt that it offered considerable advantages over traditional injection therapy, but it did not catch on, probably because it was far too large and unwieldy to expect anyone to use on a daily basis.[25]

More refined tools using similar technology – collectively known as 'artificial pancreases' – emerged in the mid-1970s. The best known of these was probably the 'Biostator'. This device could effectively automate management by continuously monitoring blood sugar levels and administering appropriate doses of insulin determined via algorithm by an integrated computer.[26]

The 'Biostator' caused a great deal of excitement when it was first launched in 1974. In 1977, George Alberti – then a professor at the University of Southampton and one of the UK's leading authorities on 'artificial pancreas' technology – was asked to return one that he had on loan from the USA. In an unprecedented move, the BDA announced that it would fund a replacement for the staggering sum of £22,000 –

around £140,000 today. The news made the front page of *Balance*, the BDA's regular member's magazine, which hopefully suggested that, one day, similar technology might be miniaturized and implanted to 'act like the real organ', removing the need for daily injections and mitigating the risk of long-term complications.[27]

In reality, however, the 'Biostator' was only ever practical for use in a hospital setting, and even there it was often unreliable. It was the size of a filing cabinet and had to be wheeled around on a trolley, meaning that those using it were effectively immobilized. Compounding this, it was prone to mechanical failure, and even when working as intended required a trained technician on hand at all times. In the end, the technology fell far short of expectations, and it was never widely used.[28]

These early experiences did, however, encourage researchers to pursue the idea of pump technology. In Paris in 1974, Gérard Slama had experimented with the concept of treating diabetes via the constant infusion of low doses of insulin. Slama had done so intravenously, which had caused practical difficulties, but his work prompted interest.[29] The healthy pancreas does, after all, operate on a principle of slow, continuous infusion. Perhaps this was the key to more 'physiological' control of blood sugar? Could the same principle be used while delivering insulin subcutaneously – just under the skin – as most injections did? Would this not sidestep most of the issues Slama had experienced?

In 1978, a paper by John Pickup answered resoundingly in the affirmative. Pickup, a diabetologist working at Guy's Hospital in London, noted excellent results in a trial using the Mill Hill infuser, the world's first successful continuous subcutaneous insulin infusion (CSII) pump (see Figure 5).[30] This had been adapted from a device used to administer drugs to experimental animals, and, unlike 'artificial pancreas' technology, made no use of feedback from any blood sugar meter – dosages were manually determined just as they would be by someone using injections. Forgoing prolonged-action varieties entirely, the Mill Hill infuser released a tiny amount of soluble insulin every few minutes – just enough to keep sugar levels stable. Before meals, a much larger dose was administered to deal with the carbohydrate eaten. This became known as the basal-bolus method – a constant flow of 'basal' (or baseline) insulin to maintain stability, supplemented with mealtime 'boluses' (single large doses) as necessary.

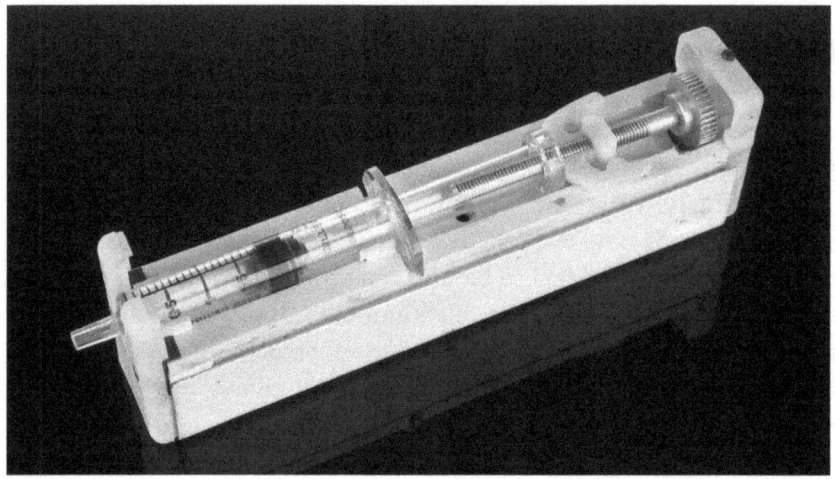

Figure 5.   *The Mill Hill infuser, 1976*

Pickup had not initially intended the Mill Hill infuser's basal-bolus strategy as anything but an experimental protocol designed to maintain 'long periods of strict metabolic control . . . to compare with "ordinary" levels of control and test the relationship between control and microvasular disease', but his paper was hugely influential.[31] Here was a device that could, apparently, quite effectively replicate 'physiological' insulin patterns without many drawbacks. Even better, it was small enough to be practically carried on the person in a discreet bag.

Soon, other studies from across the world began to demonstrate very positive outcomes using CSII technology, paving the way for the adoption of pump treatment, for some, in routine diabetes management.[32] By the mid-1980s, numerous different models existed, and, for those who could access them, they often seemed to work very well indeed.

Was this the solution to 'intensification'? In a word, no. CSII technology was certainly very impressive, but it remained out of reach of most people using insulin. The pumps were expensive, and even in countries with universal access to healthcare, the authorities were often sceptical of the cost. They were, effectively, only ever a realistic option for the wealthy until relatively recently, perhaps alongside a handful of people who acquired one by participating in a clinical trial. What was needed was a way of achieving similar results without any complex machinery. Could a comparable level of 'intensification' be achieved with simple injections?

'Self-Adjustment'

When the concept of SBGM first began to be seriously discussed in the late 1970s, early studies were optimistic that it might allow for traditional needle-and-syringe based therapy to be successfully 'intensified'. Whereas the HbA1c test could assess average blood glucose levels over a relatively long period, one of Tattersall's 1978 articles argued, accurately recorded SBGM 'not only detects poor control, but also shows how to correct it'.[33]

The results of these initial trials seemed promising. At the end of the study Tattersall was involved with, almost half of the 67 participants were reporting daily blood glucose levels 'in which no more than one ... value exceeded 10 mmol/l'. In the same issue of *The Lancet*, Peter Sönksen reported similarly positive findings, and even went so far as to claim 'that self-monitoring of blood-glucose by diabetics makes possible, for the first time, the achievement of near normoglycæmia'.[34]

Outside of the context of a clinical trial, however, this implied something rather novel. In most cases people using insulin would see a doctor only infrequently – perhaps once or twice per year – but if the results of SBGM were to be used effectively, people using insulin would have to learn to adapt their own treatment to an unprecedented extent. By the early 1980s several papers had addressed this, detailing methods for teaching 'self-adjustment' – some of them involving quite complex algorithms.[35]

One enthusiastic proponent of the concept in Britain was Anthony Knight.[36] In January 1980, he responded sharply to a critic of SBGM, who had suggested that, to ensure accuracy, blood tests should be assessed only in a laboratory.[37] 'With regard to self-monitoring by the patient', he argued, 'one fundamental question is in danger of being overlooked – who needs to know the blood glucose results?' For him, the answer was clear: 'it is the patient who needs the result and he or she needs it at the time of the test, not two or three days later ... [and to be] trained to react to abnormal patterns of results by making sensible changes'.[38]

In 1981, Knight outlined the basic premise of his approach to treatment in an article for *Balance*:

> It is a system in which a diabetic regularly measures his own blood glucose level and uses the information to control his diabetes by making appropriate adjustments to diet or insulin . . . [the objective being] for the diabetic to become his own laboratory technician, dietitian and doctor in the day to day management of diabetes.[39]

The principle of 'self-adjustment', in this conception, existed as a tool by which recurrent episodes of high or low blood sugar could be analysed and addressed within the familiar framework of treatment. This continued to occur within the context of the paternalistic tradition – changes by the insulin user were expected to be made extremely cautiously, and always according to precise instructions:

> Treatment changes are made in small steps such as shifting 10 grams of carbohydrate or changing an insulin dose by two or four units . . . Medical guidance is essential for this stage . . . Changes in treatment should be small, should only be made every two to three days and should be made to correct an identified pattern of abnormality.[40]

In the early 1980s, asking those with diabetes to take even this level of responsibility was controversial. Knight, for example, defensively insisted that 'self-adjustment' was 'not a ploy to get diabetics to look after themselves to make the busy diabetic clinics more manageable'.[41] How many actually did begin to use SBGM is not clear, but many certainly did not. Half a decade later, physician-patient writing duo Judith M. Steel and Margaret Dunn reported that many people using insulin continued to refuse to make any adjustments to their treatment, claiming that they were 'told never to mess about with the insulin', and that they 'should be far too scared' to do so.[42]

Nonetheless, almost every article published in this period emphasized the popularity of SBGM amongst those who did begin incorporating it into their management. As one 1980 paper put it, 'the assumption by patients of the responsibility for management of their own disease tends to break the pattern of nihilism and frustration often found in both patient and physician'.[43] Philip Newick, for example, began to use a meter in the mid-1980s, and reflects that it 'changed his life'. Being able to access accurate information about his blood sugar levels allowed him to predict and manage fluctuations much more readily, particularly when it came to avoiding hypoglycaemic episodes, while the control it implied also made his treatment much more psychologically bearable.[44]

While all agreed that almost everyone who used SBGM preferred it to urine tests, however, the notion that it could be instrumental in achieving 'intensification' was much more hotly debated. One 1982 paper argued, for example, that, in those studies that did show reductions in overall blood sugar levels, the results could be attributed primarily to motivated participants and close medical supervision, with SBGM offering 'no improvement in control over intensive attention and conventional urine monitoring'.[45]

This, perhaps, reflected the inherent limitations of 'self-adjustment' as Knight and others in the early 1980s conceived it. The vast majority of people using insulin at this time remained on fixed insulin regimens – those with T1DM typically took two shots per day, often a mixture of soluble and NPH or Lente. While doses and diet could be adjusted, this remained an unwieldy system that could not effectively adapt to inevitable day-to-day fluctuations. A correction made on one day could be wholly inappropriate for the next.

One important innovation in treatment that went beyond the kind of cautious strategies advocated by early advocates of SBGM, however, was the introduction of multiple daily injection (MDI) treatment. It is actually quite bizarre that such an approach seemed as novel as it did to some in the 1980s. While he thought it impractical, R.D. Lawrence had suggested as early as 1949 that more stable blood sugar might be achieved 'by giving small doses of insulin four or five times in 24 hours'.[46]

In the mid-1980s, however, a great deal of work appeared on the topic.[47] This was almost certainly at least partially prompted by the development of so-called insulin pens. Novo Nordisk's 1985 'Novopen' is often credited as the first example of this technology, but this is actually incorrect.[48] The concept had first been implemented by a group

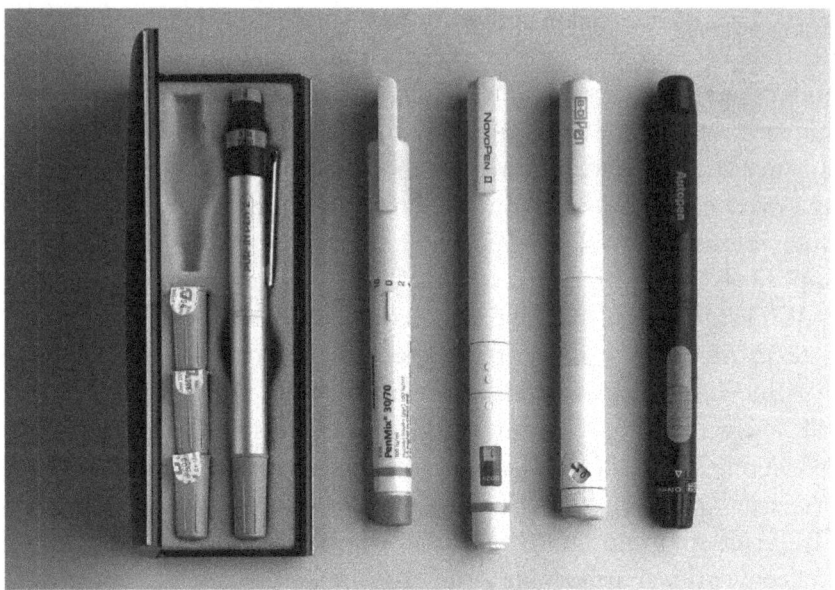

Figure 6.   *A selection of insulin pens, 1985–1993*

of doctors in Glasgow in the late 1970s after Sheila Reith, one of their number, was struck by the unwieldiness of the traditional glass and steel syringes her daughter was asked to use following a diagnosis of T1DM.[49] From 1983, their design was commercially available as the 'Penject', but it could not compete with later models released by much larger corporations (see Figure 6).[50]

Insulin pens, as their name suggests, resemble large writing pens. After they are loaded with a specially designed cartridge of insulin, they can deliver an easily-adjusted shot at the touch of a button. Pens made the prospect of administering insulin outside of the home – a vital part of MDI – significantly less of a trial because they were much more discreet than a needle and syringe. Injections could now be given in a matter of seconds at a restaurant table, for example, and few present would even notice.

MDI took many forms – the term itself simply referred to any regimen that involved more than the standard two daily injections. One of the most important opportunities it presented, however, was making the kind of basal-bolus principles used in CSII feasible for those without pumps – the same thing being achieved via creative pharmaceutical, rather than mechanical, manipulation. If longer-acting varieties like NPH or Lente were used to establish a stable 'basal' rate, could soluble insulin – with its fairly intense rate of action – not be used exclusively for 'boluses' at mealtimes?[51]

One of the first studies to discuss basal-bolus MDI treatment was actually something of an outlier, published in *Diabetes Care* in early 1978. The paper described a style of management in which insulin was delivered four times per day – one dose of NPH at bedtime alongside soluble insulin before each main meal, individually adjusted as necessary based on SBGM testing.[52] While the authors did not believe that 'all patients with insulin-dependent diabetes require multiple injections of insulin', they did claim impressive results, and suggested that the approach 'usually moves the degree of control closer to the optimum'.[53]

Like SBGM-facilitated 'self-adjustment' more generally, MDI using insulin pens quickly became highly popular amongst those using insulin. This is not particularly surprising. One of the greatest advantages of the strategy was the quality-of-life benefits that it permitted – so long as an adequate 'basal' rate was in place, meals could be taken at any time, and insulin could be adjusted to compensate for more variety. Nonetheless, like SBGM before it, some promising initial results in early, small-scale clinical trials were soon followed by research claiming it did little to reduce overall blood sugar.[54] By the 1990s, the growing consensus was that, if it did have any effect, it was a very modest one.[55]

'It should not be assumed', one 1989 article suggested, 'that glycaemic control . . . will improve as a direct result of an increased number of daily injections'.[56] Two years later, another even claimed that it might lead to deterioration – that by allowing for more flexibility in food there was 'greater scope for dietary indiscretion'.[57] The BDA's Judith North – who lived with T1DM – alluded to these critiques in a chapter that she wrote for the 1991 edited volume *The Technology of Diabetes Care*. While she thought that MDI, which she had switched to a few years prior, had 'brought a much improved quality of life and valuable flexibility', she admitted that, 'probably', it implied 'no better "control"'.[58]

By the beginning of the 1990s, new methods of delivering insulin had dramatically changed the landscape of diabetes management, and those who used 'self-adjusted' frameworks seemed in almost all cases to much prefer them to traditional fixed-injection regimens. While these novel approaches did, on paper, seem more 'physiological' than their predecessors, however, with the exception of expensive CSII treatment, there seemed precious little real evidence that they produced better outcomes from a clinical perspective.

But how much did any of this actually matter? The idea that persistently high blood sugar was directly connected to the development of long-term complications was by this point broadly accepted by the vast majority of medical professionals, but this was grounded primarily in theoretical arguments alongside a few very small-scale studies.[59] Where was the hard proof?

The Diabetes Control and Complications Trial (DCCT)

In 1993, a study published in the *New England Journal of Medicine* made the uncertain efficacy of most contemporary treatment strategies seem especially troubling. The DCCT was one of the most significant diabetes-related clinical trials of the late twentieth century, and it proved beyond reasonable doubt both that high blood sugar could contribute to the development of long-term complications, and that 'intensive' management could mitigate this risk.[60]

The trial, which ran from 1983, recruited 1,441 participants with T1DM from twenty-nine separate institutions across North America. Half were placed into a control group and were treated according to the kind of static, prescribed framework typical of contemporary diabetes management. They took one or two daily injections of insulin, were encouraged to test their urine and/or blood sugar occasionally, and received basic training on nutrition and exercise. Conversely, the remainder were placed on an experimental 'intensive' regimen entailing '[the] administration of insulin three or more times daily by injection or an external pump . . . [with] dosage adjusted according to the results of self-monitoring of blood glucose performed at least four times per day, dietary intake, and anticipated exercise'.[61]

The control group's treatment goals would have been familiar to anyone undergoing insulin therapy in the mid-twentieth century. They attended a clinical appointment once every three months, but had no specific blood sugar targets. Instead, they aimed to ensure 'the absence of symptoms attributable to glycosuria or hyperglycaemia; the absence of ketonuria; the maintenance of normal growth, development, and ideal body weight; and freedom from severe or frequent hypoglycaemia'. By contrast, the expectations placed on the 'intensive' group were considerably more demanding. They 'visited their study centre each month and were contacted even more frequently by telephone to review and adjust their regimens':

> The goals of intensive therapy included preprandial [before eating] blood glucose concentrations between 70 and 120 mg per deciliter (3.9 and 6.7 mmol per liter), postprandial [after eating] concentrations of less than 180 mg per deciliter (10 mmol per liter), a weekly 3-a.m. measurement greater than 65 mg per deciliter (3.6 mmol per liter), and hemoglobin A1c (glycosylated hemoglobin) [HbA1C], measured monthly, within the normal range (less than 6.05 percent).[62]

Clearly, these participants were being managed very differently, and their treatment was continuously adapted according to planned diet and lifestyle. This, however, asked a great deal more labour of everyone involved – more injections, more blood sugar tests, and frequent contact with professionals, who in turn were expected to use the information they were given to quickly draw up revised prescriptions where necessary. Nonetheless, by the strict clinical criteria set out by those running the trial, 'intensive' treatment seemed a resounding success – it was so successful, in fact, that the study was terminated earlier than planned on ethical grounds after it became obvious that the two groups were experiencing stark differences in outcomes.

Those undergoing 'intensive' treatment had, on average, HbA1c levels of around 7%, while in the control group the figure trended at approximately 9%. Quarterly assessments also found that, when given spot blood sugar tests, the mean value amongst the former was 8.6 mmol/l, but in the latter 12.8 mmol/l. Most importantly, 'intensive' management seemed to dramatically reduce the likelihood of developing long-term complications, showing a 63% reduction in sustained retinopathy (eye damage), a 54% reduction in advanced nephropathy (kidney disease), and a 60% reduction in neuropathy (nerve damage) over the control group. The only observed shortcoming appeared to be an approximately tripled risk of severe hypoglycaemia.

In their concluding discussion, the DCCT Research Group endorsed the widespread adoption of 'intensive' therapy amongst all but the most incapable (or unwilling) people with insulin-dependent diabetes – that is, T1DM. This, however, they acknowledged, posed a significant challenge. The labour and resources required to so closely supervise so many people were significant, and might be prohibitively expensive:

> Intensive therapy was successfully carried out in the present trial by an expert team of diabetologists, nurses, dieticians, and behavioural specialists, and the time, effort, and cost required were considerable. Because the resources needed are not widely available, new strategies are needed to adapt methods of intensive treatment for use in the general community at less cost and effort.[63]

The DCCT's organizers understood the respective roles of healthcare professionals and their patients in the business of diabetes management in fundamentally paternalistic terms. In the highly micromanaged 'intensive' group, this was especially important. Complete passivity was not only considered acceptable, it was framed as wholly desirable. If anything, the patient as a subjective entity, living as they did in the messy, unpredictable context of day-to-day life, posed a threat to strict control – it was a chaotic force that needed to be restrained.

When the DCCT concluded, the problems with this approach became clear. British diabetologist Stephanie Amiel, for example, succinctly points out how, once they left the study, the clinical improvements seen in its participants quickly disappeared:

> [In the] DCCT, they had 2.8 patients per researcher . . . The patients on the 'intensive' arm were contacted weekly by the diabetes nurse. They reported their blood glucose results and she told them what to do for the next week. So, it was all run by [the clinicians] and . . . at the end of nine years of intensive therapy, when they stopped DCCT prematurely, the patients in it could not sustain the HbA1c because they'd learned nothing in nine years of being told what to do![64]

It is not surprising that those in the 'intensive' group of the DCCT struggled once the study ended. The project's whole approach to achieving its goals had been to provide continuous bespoke guidance that accounted as best as possible for potentially disruptive factors. It had not taught the individuals involved anything about their own management. Without the support network that had previously sustained them, they had no way to continue.

Whatever criticisms can be made of the DCCT's approach to 'intensive' treatment, however, it did work on its own terms. Nonetheless, while the strategy was, in this sense, sound, medical practice does not occur in a vacuum. Even as they congratulated themselves on their results, the organizers of the study must have known that they had also highlighted its practical unfeasibility at a larger scale. The sheer amount of resources necessary to roll out the principle more broadly seemed almost insurmountable.

Düsseldorf

Across the Atlantic, however, an alternative approach had begun to coalesce some years before, pioneered by members of the European Association for the Study of Diabetes (EASD). In 1983 – the same year that the DCCT began – representatives of the organization's Diabetes Education Study Group (DESG), led by Michael Berger, reported that they had observed overwhelmingly positive results after implementing a pioneering approach to management in Düsseldorf in 1978.

While the German group agreed that strict control of blood sugar likely reduced the risk of long-term complications, and thought that this should be one of the primary goals of insulin therapy, the strategy they employed to facilitate it was almost diametrically opposed to that of the DCCT. Instead of trying to control almost every aspect of their patients' lives, Berger and his colleagues sought to enlist them into their own care by utilizing what he called a Diabetes Teaching and Training Programme (DTTP). This complex, five-day initiative promoted basal-bolus principles, but more than this it also covered almost every aspect of practical diabetes management and the theoretical relationship between medication, food, and lifestyle. If someone could recognize shortcomings in their own management style and respond accordingly, they could, so the thinking went, bring blood sugar fluctuations back under control without having to wait for a doctor or other healthcare professional to advise them.

The premise of extensive patient education was by this point nothing new, and the original DTTP implemented by Berger was adapted from another devised by Geneva's Jean-Philippe Assal.[65] In Düsseldorf, however, the concept was taken in bold new directions, and in the process meaningfully reconstituted the role of the layperson in management to an unprecedented extent.

The German-speaking world must have seemed an unlikely place for the birth of what ended up becoming a revolutionary approach to insulin therapy. Following the Second World War, orthodox diabetes management there remained characteristically intensely paternalistic. Represented by figures like West Germany's Ferdinand Bertram and the East's Gerhardt Katsch, this school emphasized strictly prescribed diets, expected absolute obedience, and dissuaded people from engaging with their treatment any more than was absolutely necessary.[66]

Former EASD president and one-time colleague of Berger in the DESG, Victor Jörgens, suggests that one reason for this was that during the Second World War many of Germany's most talented practitioners either fled the country or were killed by the Nazi regime, leaving only a 'weak remnant from the second league . . . to write textbooks and to organize diabetology'.[67] The field, as a result, became stagnant and unimaginative.

This culture persisted to some extent well into the 1960s and 1970s. By the time Berger began working there in 1978, Düsseldorf was known as a specialist diabetes centre with dedicated wards and outpatient provision, but the new arrivals later noted how conservative the institution seemed to be regardless:

> Previously, diabetes education was delivered on an unstructured individual basis, with regular metabolic self-monitoring (urine and/or blood glucose testing) being taught to selected patients only. Patients were not advised to alter their insulin dosage without contacting a physician and major emphasis was placed upon a rigid dietary regimen.

Berger, however, rejected this 'cumbersome therapeutic regimen' and criticized the scattered, inconsistent efforts at patient education that characterized it, arguing that this vital element of treatment had been 'neglected during recent decades ... [and] delivered mainly on an unstructured basis without any systematic attempt to evaluate its long-term effectiveness'.[68]

New technologies like pumps, pens, the HbA1c test, and SBGM equipment were all useful, but – implicitly – alone they were useless. If people with diabetes did not understand their own treatment intimately and remained essentially reliant upon largely static prescriptions written up by a distant physician who might, at best, tinker with the formula every few months, how could they possibly be expected to maintain stable blood sugar levels? Even where they did incorporate some modest scope for 'self-adjustment', most contemporary physicians tried to super-impose this over a relatively traditional framework. As late as 1991, for example, Harry Keen wrote that, when it came to the results of SBGM:

> The crucial decision the patient has to make, once or several times a day, is how much insulin to inject: the usual, prearranged dose if the blood glucose is correct, rather more if it is too high, and less if it is too low, modified perhaps if there is some anticipated change in mealtimes, meal size or physical exertion.[69]

In the chaotic context of real life, however, the Düsseldorf group implicitly suggested, any kind of 'prearranged dose' simply made no sense – except perhaps as an initial jumping-off point. This kind of rigidity, they argued, was the key to explaining why the results of some prior SBGM trials had been disappointing:

> These authors failed to show any improvement in glycosylated haemo-globin levels persisting beyond 6 months after the end of the teaching programme. Their failure might have been partly due to the fact that the method of diabetes education used was quite different from that used here, and also to the rigid rules for metabolic self-monitoring which had been imposed on their patients. The results of the present study underline the advantage of allowing the 'educated' patient to choose his/her individual schedules and methods of self-monitoring.[70]

To Berger and his colleagues, 'education' came to represent something very different to the extremely cautious programmes seen in other locations. The DTTP implemented in Düsseldorf did not simply train participants to respond in an 'approved' way to the results of self-testing. Instead, it created an officially sanctioned, highly empowered role for them within treatment.[71]

This approach soon seemed promising. Five years after taking over, those who had attended their DTTP seemed to have much more stable blood sugar levels than they ever had previously. Following a full evaluation twelve months after the course, average HbA1c levels amongst those who had taken part had fallen by 1.6%. Additionally, they were also significantly more self-reliant, and were admitted to hospital 90% less frequently. A slight rise in cases of severe hypoglycaemia was noted, but this was modest and the authors did not consider it any real cause for concern.[72]

Berger and his team clearly believed that their approach offered significant advantages at little cost. Their paper concluded by recommending that the kind of DTTP used in Düsseldorf – and, from 1981, Vienna – be offered to almost all people with T1DM, and once again it emphasized the value of patient participation in management:

> All patients, irrespective of their educational status or intelligence, should be trained for self-management of their diabetes. Permanent improvement of metabolic control can be maintained only by daily self-monitoring of blood glucose or urine sugar and appropriate adaptation of the treatment by patients themselves.[73]

A few years later, in 1987, the group published another study showing that they had successfully replicated their results in a further trial in Bucharest. The exact same DTTP was implemented – two Romanian physicians and two nurses visited Düsseldorf in 1984 and were trained to deliver the course exactly as it was taught in Germany and Austria. Once again, the results were positive.

Berger's approach did not only deliver clinical benefits. Unlike those in the DCCT, people who had attended his DTTP also reported great improvements in their quality of life. In particular, they appreciated being able to break the monotonous routine of fixed diets and scheduled mealtimes. Knowing how to more confidently manipulate their insulin, it had become possible to introduce more variety to their food, to eat at different times, and even sometimes to skip a meal entirely if they were not hungry.

Even the earliest of the Düsseldorf group's articles mention the DTTP's potential to allow the traditionally rigid approach to diet to be somewhat relaxed, but after their experience in Romania they felt confident enough to take this claim even further. With a sufficiently educated body of motivated patients willing to give themselves sufficient injections, there was almost no limit to the flexibility on offer:

> Under the condition that insulin treatment is intensified, patients may liberalise their diet. This includes variation of amount and timing of carbohydrate intake, skipping of meals altogether and a prudent consumption of sucrose and sucrose containing nutrients. The more liberalised the diet becomes, the more frequent measurements of blood glucose and the more frequent injections of regular insulin and immediate adaptations of insulin dosage will be necessary to keep glycaemia in optimal control.[74]

Unsurprisingly, this was enthusiastically embraced by their patients. On entry to the 1987 study, almost every single participant had taken, on average, two injections per day – entirely typical of the time. Two years later, a full 62% used some variation of an MDI approach – often adopting basal-bolus principles – that involved three or more shots each day with regular ad hoc adjustments to dosage to account for variations in diet and lifestyle.[75]

In practice, Berger had turned insulin therapy on its head. Rather than attempting to fit his patients' lives around an imposed pattern of medication, he had given them the tools to safely modify their treatment regimen as required by their own personal life-context and, perhaps more importantly, the explicit permission to do so.

This might have seemed revolutionary at the time, but Berger had actually taken the concept from Karl Stolte, a Breslau – now Wrocław – physician often unfairly ignored by the historical literature. Working as a paediatrician in 1928, Stolte had been emotionally moved when he witnessed a group of children sadly crowding around the bed of one of their ward-mates on his birthday. The child's mother had brought in a cake and other sweets, but the group – each of whom had been diagnosed with diabetes – were forbidden from partaking. They could only look on enviously.

Stolte vowed that from that day forward he would ensure that, on his ward, even children with diabetes would be allowed a cake on their birthday, and would be allowed to share it with their friends. Shortly after, he made good on his promise, and gave a little extra insulin beforehand to balance out the additional carbohydrate. His colleagues were anxious, but to everyone's surprise the children's urine tests showed no deterioration. If anything, they were passing less sugar than usual.

Afterwards, Stolte began to take a then-radical approach to practice. He allowed his patients to eat more or less whatever they liked, and taught them to take a tailored amount of soluble insulin before meals. This was a clear precursor to basal-bolus thinking (though of course initially minus the 'basal' element), but Stolte clearly had more than pure clinical efficacy in mind. 'Diabetics', he later wrote, 'should not be treated like laboratory animals that receive an exact amount of food every day.'[76]

Stolte was largely dismissed in his own time, but when Berger rediscovered his ideas years later, he was highly impressed, and he spoke highly of his predecessor – even going so far as to write a book about him.[77] The affinity he felt is not surprising. Both men shared a subtle insight that informed their approach to medicine – that requiring insulin does not in itself make someone 'sick'. Everyone requires insulin. Some of us just have to get it into our bloodstream in a more roundabout way than others. People with diabetes fulfil few if any of the characteristics of the 'patient' in their day-to-day lives. For most of the time, they live much as anyone else does. Insulin does not cure anything. It is not a medicine. It exists in everyone. For some, whose bodies have ceased to make enough, or who have become resistant to what they do make, injections provide an external source, but they are no more *ill* on any given day than anyone else.

The predisposition of the medical profession to assign people using insulin permanent 'patient-status', and to sternly prescribe fixed treatment regimens as it would for any other condition, has always been counterproductive. Insulin needs do not remain the same every day, even where a standardized diet and lifestyle are imposed. Physicians have always acknowledged this reality, but until relatively recently there is little evidence that they collectively reflected on its deeper implications to any meaningful extent. How could a rigid treatment schedule hope to maintain stable blood sugar levels – much less consistently 'normalized' ones? The entire premise, viewed in this way, seems utterly ridiculous.

The Düsseldorf group, however, understood this. Insulin therapy, its members argued, is simply the replacement of a hormone. So long as that hormone is delivered appropriately, nothing else matters. There should be no practical need to insist on restricted diets if treatment can be manipulated effectively instead. There is no morality in taking more or less insulin, or in engaging in unnecessary self-deprivation.

By encouraging autonomy in insulin users and jettisoning the prescriptivism that so often defined contemporary management, treatment could satisfy the 'success' criteria of all parties. Average blood sugar levels could be reduced. Pressure on hospitals could be brought down. Dramatic improvements to quality of life could be won. All of this could happen because people were able, in all but the direst situations, to look after themselves.

Conflict between the organizers of the DCCT and the Düsseldorf group was almost inevitable. The former believed that 'intensive' control could be achieved only by strict professional micromanagement, while the latter felt that the whole premise was an artefact of counterproductive medical authoritarianism, and that the opposite should be done – insulin users should be provided comprehensive education and then left well alone outside of occasional check-ups unless they specifically requested help. Amiel remembers some of the blazing arguments that Berger, often alongside his long-term collaborator and wife Ingrid Mühlhauser, would have with their representatives at conferences:

> He was quite a showman, Michael, and he would stand in meetings and he would throw his arms about and say 'T1DM is an insulin deficiency disease and you treat it with replacing insulin, not with diet!' . . . And there were some wonderful meetings where the Americans, the DCCT

people, would be pitted against Ingrid, who was quite amazing, and Michael, and she would wipe the floor with them! She would wipe the floor with anybody![78]

With the exception of the German-speaking world and some parts of the Eastern Bloc, however, the Düsseldorf group had limited initial influence. Berger's major textbook, *Praxis der Insulintherapie*, was, for example, never translated from German.[79] One of the only works that did enjoy wider circulation was *Funktionelle Insulintherapie*, a 1987 book by Vienna's Kinga Howorka, which received an English translation in 1991.[80] This provided what was, at the time, probably one of the most comprehensive English language descriptions of the Düsseldorf approach, which she referred to as functional insulin therapy (FIT), as it stood at the beginning of the 1990s.[81] 'The goal of FIT', she declared, 'is to adapt the therapy to the life circumstances of the patient.'[82] The trick to doing this consistently and effectively, she went on to argue, was through the use of SBGM-adapted 'basal bolus' therapy, implemented by either CSII or MDI. Importantly, whereas most were content to leave it implicit, she clearly highlighted the absurdity of any 'pre-arranged' insulin dose in the context of her practice:

> There are some characteristics, however, which enable one to distinguish whether true functional insulin treatment is being carried out. For example, one needs only to ask an insulin-dependent patient, 'What do you inject?' This usually means what kind of insulin, how much, and what diet has been prescribed. If the patient answers without hesitation, 'X units of this insulin plus Y units of that insulin in the morning and X units of this plus Z units of that in the evening', one can be sure that it is not FIT. A patient who is using FIT cannot answer this question so easily. He would probably answer, 'It depends.'[83]

Funktionelle Insulintherapie received an initially lukewarm reception in the English-speaking world. When he reviewed the second translated edition in 1997, Keen was not entirely hostile. He even remarked that the strategy employed by Howorka strongly resembled the approach taken by Lawrence to his own treatment. Nonetheless, he was clearly a little uncertain about the implicit message of the work. 'This salty little book', he suggested, exhibited a 'common sense Messianism', but, while stimulating, 'should not be taken undiluted'.[84]

Keen was very much a representative of the old guard of British diabetology by 1997, and his cautious response to Howorka reflects this. A few years before, following the publication of the results of the DCCT, he had written an opinion piece in *Balance* that expressed concern about the implications of the study for the mental welfare of those using insulin:

> It was a marvellous study but I have mixed feelings about it. I start from the assumption that the one thing people with diabetes want is to not be diabetic; anything which brings the condition more firmly to their attention or involves them in more activity is something which they may greet with less than wild enthusiasm.[85]

It is not surprising that Keen remained sceptical of the Düsseldorf approach. For all of its promises of liberation from restriction, it did ask a lot of its adherents. If the DCCT demanded too much patient engage-ment, then this was clearly even worse! People with diabetes would never be able to escape their condition! This, he predicted, would be a significant barrier to its adoption. Keen was, however, wrong on this point. The day-to-day reality of diabetes cannot be escaped whatever style of regimen is used. However, his attitude provides a clue as to the kind of subjective thinking that prompted his suspicion of FIT. Implicitly, the idea that doctors might give up even more control over their patients seemed instinctively troubling.

In his foreword to the first edition of *Funktionelle Insulintherapie*, Berger directly alluded to this. The Düsseldorf strategy was not merely the product of scientific and medical innovation, he argued, but was part of a much broader cultural debate about individual autonomy that had prompted widespread critical engagement with social power structures. Medicine, of course, was one of the clearest examples:

> This new orientation in the therapy of type 1 (insulin-dependent) dia-betes mellitus cannot solely be attributed to the latest scientific and technological developments and discoveries. It also coincides with a particular sociocultural trend of recent years, namely, the dissolution

of traditional authoritarian structures. There could be practically no more paternalistic and dependent relationship than that of the traditional relationship of the (chronically ill) type 1 diabetic patient and his physician. The collapse of this completely inefficient, even inhuman and quite unjustifiable subordination and the emancipation of the patient within a cooperative partnership with his physician represents a prerequisite for the successful performance of intensive insulin therapy by the patient.[86]

While Berger himself was, like most of those advocating for this style of management, very enthusiastic about the promise of a more equitable relationship between patient and practitioner, many doctors felt threatened by the idea. For some, certainly, this was largely a jealous defence of professional privilege and the social prestige – and pay cheque – that went with it. But there was also a more existential question presented here: if someone with diabetes was able to largely look after themselves, and in many cases achieve better results, then what did that say about the role of the doctor?

Proto-Intensification

For most of the twentieth century, debates between physicians about the 'correct' way to manage diabetes tended to adopt a distinctly top-down perspective. How should insulin and diet be prescribed? What level of 'control' is appropriate? Those actually using insulin, however, were usually completely excluded from meaningful participation in this discourse.

In the context of insulin therapy, however, physicians were (and remain) essentially paper tigers. If their patients choose to ignore them, they are powerless to do much about it. It is the people with diabetes themselves who make the final decision on medication, diet, and lifestyle. Since its first use in 1922, insulin has unavoidably thrust control into the hands of the person using it, whether or not either they or their doctor like it.

Without anyone to stop them, some people using insulin have always rejected orthodoxy. In a 1980 volume of *Diabetes Care*, for example, Bernstein – one of the first to use an SBGM machine in his home

– described how, after many years studying his condition, he had developed his own approach to treatment based on three daily shots with regular corrective doses of soluble insulin where necessary.[87] His self-designed regimen strongly resembled later MDI frameworks proposed by physicians, and he claimed to have achieved almost consistently 'normal' blood sugar levels by using it. Tellingly, however, Bernstein initially struggled to publish his research. In the end, he went so far as to retrain as a physician to ensure that his work would be taken seriously. 'I never wanted to be a doctor', he claimed in a 1988 interview, 'but I had to become one to gain credibility.'[88]

Independently-minded figures had, however, been doing similar long before. In 1925, for example, 13-year-old Jack Eastwood was diagnosed with 'severe' diabetes mellitus and quickly put onto a strict regimen typical of the 1920s. Writing about his life in an article for the *BMJ* in 1986, many decades later, he reflected on the limited education he was given in hospital. In short, it 'was confined to a clear understanding of how important it was to keep to the rules, and to early recognition of the symptoms of an overdose of insulin':

All of my food was weighed, and no excesses were allowed at all ... subcutaneous injections of soluble insulin were given before breakfast and supper each day ... about 20 U with the carbohydrate content of my food amounting to about 25 g.[89]

As he entered adolescence and won a scholarship to St Paul's Public School in London, Eastwood began to be troubled by the claustrophobia of such an inflexible and restrictive routine. He was particularly frustrated by the idea that it might hinder his pursuit of an active social life, or indeed his ability to do anything with spontaneity – a feeling that only intensified when he moved to Oxford to attend university.

Determined to master his condition, he assiduously compiled notes detailing his consumption of food and insulin, the results of urine tests, and any miscellaneous observations he felt were relevant. The relative ease of life at Oxford – he found the time to play golf 'nearly every afternoon' (!) – gave Eastwood the opportunity to analyse his records and develop, with considerable trial and error, a bespoke approach to treatment based on constant adaptation, adjustment, and self-assessment:

In brief summary, its distinctive feature is that I inject insulin at every mealtime and vary the dose according to the food eaten, instead of basing the treatment on a fixed dose of insulin each morning and then trying to adjust my diet, exercise, etc, to this throughout the next 12 or 24 hours . . . eating almost anything I wanted . . . due allowance being made for what I expected to be doing during the next few hours.[90]

By adopting this (at the time highly unorthodox) style of treatment, Eastwood was able to adapt to the changing demands of his lifestyle and, as a result, was able to escape the monotony experienced by many of his contemporaries. He went on to enjoy a successful career as a schoolteacher, eventually becoming headmaster of Ermysted's Grammar School in Skipton, North Yorkshire. Despite rejecting the strict routine expected of those using insulin, he reported consistently satisfactory urine tests and was happy to have developed no noticeable complications at the time of writing.

Eastwood died in 1987 at the age of 75.[91] While this lends his optimistic 1986 paper an undeniably melancholic air, he was nevertheless extraordinarily long-lived for someone who had started using insulin in the 1920s. Like Bernstein, he had also developed and documented a fully realized alternative framework of management based on meticulous self-observation and analysis that utilized MDI principles.[92] Importantly, however, he had done so in the 1930s!

Eastwood did not visit a doctor about his diabetes after 1935. Had he decided to, he might have found them less than enthralled. Many would almost certainly have considered his rejection of professional prescription dangerously irresponsible. He himself well understood this, directly acknowledging that some might feel that his 'attitude entailed an unwarrantable risk . . . that [he] was lucky not to suffer in consequence of'.[93]

In 1980, an article in *Balance* by Charles Fletcher – at the time a household name in Britain as the presenter of pioneering TV documentary series *Your Life in Their Hands* – outlined a very similar strategy to Eastwood. Always a critic of undue medical authoritarianism, he provocatively declared that 'it isn't necessary to be a doctor to do this sort of day by day adjustment'.[94] Despite his status as a professional, Fletcher received sharp criticism from several readers of *Balance* – most of them laypeople living with diabetes. One correspondent, for example, doubted 'whether he has shown us a good example of diabetic control', while another claimed quite firmly that 'the attitude of the article was one that would be viewed with horror by most of those looking after diabetics'.[95] To them, he seemed to be acting thoroughly irresponsibly – they had, after all, been warned of the importance of faithful obedience their whole life.

These examples highlight the entrenched medical authoritarianism that permeated attitudes towards diabetes management at the beginning of the 1980s. Even amongst the patient-body, the suggestion that anyone using insulin should make changes to treatment without the approval of their doctor was met with great suspicion.

By the late 1980s, however, something had begun to change. Vocal criticisms of healthcare professionals began to appear much more frequently in magazines like *Balance*. Even established publications like the *BMJ* appeared to have become at least somewhat less dismissive of laypeople. Eastwood – a retired schoolteacher with no formal medical training whatsoever – was, after all, able to have them publish his article. What had happened to make this possible, and what did it mean for the future of insulin therapy?

Chapter 4

Subjectivity, Paternalism, Neoliberalism, 1993–2002

In 2002, the *BMJ* published a paper titled 'Training in flexible, intensive insulin management to enable dietary freedom in people with type 1 diabetes'. This outlined an experimental DTTP that had been trialled in London and the north of England with funds from Diabetes UK just after the millennium. The Dose Adjustment for Normal Eating (DAFNE) programme, as its organizers christened it, promoted a radically different approach to treatment than had ever previously been attempted at scale in Britain.

The authors of the study explained their rationale. It was still uncommon, they argued, for people using insulin in the UK to meaningfully adjust their own dosages or otherwise 'intensify' their management, and they rarely maintained the 'degree of glycaemic control known to be ideal'.[1] The DCCT, however, had demonstrated quite clearly this put them at significant risk of long-term complications, but had also highlighted how tricky it could be to do much about it, especially where budgets were limited.

DAFNE, they suggested, offered a potential solution. Forgoing traditional fixed diets, it understood T1DM as 'an insulin deficiency disorder, best managed by insulin replacement as necessary and not by dietary manipulation to match prescribed insulin'. It sought, as a result, to teach people to adapt their own treatment as necessary independently of professional instruction.

Recruiting 169 participants, each of them considered to have 'moderate or poor glycaemic control and quality of life', DAFNE's DTTP utilized an intensive five-day training course in which they were trained to adjust their insulin doses autonomously, based on the food they ate and what they planned to do that day, making use of group activities to ensure that they developed not only a theoretical understanding of the principles at work, but also the confidence to implement them in real-world scenarios.

In order to assess the effectiveness of the programme, the cohort was split into two groups. Following an initial assessment, both attended identical courses, but one went to theirs right away while the other was delayed until later in the year. After six months – before the second group received their training – each participant was asked to attend a review appointment. The results were highly promising. The delayed group had an average HbA1c level of 9.4% – little changed from the 9.3% average they had displayed at the beginning. In those who had received their training immediately, however, the figure had dropped from 9.4% to 8.4%.

Furthermore, this appeared to have been accomplished while promoting improved quality of life. At both appointments, participants were asked to score the impact of diabetes on their quality of life on a scale of –9 (most negative) to +9 (most positive). At sign-up, the group that went on to attend DAFNE right away averaged out at –2 generally, and –4.8 when the question was refined to focus on dietary freedom specifically. By the time of the six-month review, however, these figures had fallen to –1.6 and –1.8 respectively.

The programme appeared to have been a resounding success by almost everyone's standards. While some people with T1DM might, the authors admitted, prefer 'a simpler regimen with routine meal timing and fewer injections', the study set a highly influential benchmark for diabetes care.[2] In the following years, DAFNE expanded significantly, and today DTTPs across the globe continue to recognize it as a major influence.[3] Follow-up studies evaluating its long-term effectiveness have also been generally positive.[4]

If all of this seems oddly familiar, it is with good reason. When the DAFNE study was published in 2002, it seemed quite revolutionary. However, it was in reality nothing new. The entire DTTP had, in fact, been translated almost word for word from the Düsseldorf programme, which had by then been running for over two decades.[5]

Michael Berger and his colleagues must have been surprised when a contingent of British physicians arrived in Germany in the late 1990s, keen to observe the approach to insulin therapy practised there. He had won few friends in the UK, where his outspoken rejection of orthodox, physician-led treatment had previously been largely dismissed. What, he might have wondered, had changed?

The emergence of 'patient-led' treatment in turn-of-the-millennium Britain may at first glance seem puzzling. Why now, and not years prior? In order to understand why this occurred, it is necessary to consider the broader context of contemporary diabetes management. DAFNE's appeal stemmed from a complex constellation of social, professional, and political factors.

As ever, people with diabetes were far from passive actors when it came to cultivating the environment in which DAFNE emerged. One of the programme's great deviations from traditional frameworks of care was in its attention to quality-of-life factors, and its acknowledgement that clinicians and their patients may have different priorities in treatment. This reflection on subjectivity was, as we shall see, highly important. With this in mind, it seems an appropriate place to begin our discussion. How did subjectivity shape the experience of insulin therapy, how was it expressed by the closing decades of the twentieth century, and what were the implications for all parties involved?

Subjectivity

What, exactly, is meant by 'subjectivity', however? The term is often used somewhat loosely, and scholars continue to disagree about its precise character. How, then, should we define it? For the purposes of this book, I use it in a relatively uncontroversial, broad sense. Subjectivity, as I understand it, refers to how people perceive themselves, give meaning to their lives, and interact with the world around them, and also the way in which this internal landscape shapes their personal value systems while informing their everyday decisions.[6] This, of course, is not disconnected from the material world. Subjectivity is shaped by external pressures both physical and ideological. As anthropologist

Tanya Luhrmann puts it, it is the '*emotional* experience of a *political* subject', intimately bound up with the frameworks of power and authority that define our societies.[7]

In healthcare, subjectivity has deep relevance – particularly when it comes to assessing concepts of 'success' and 'failure'. A clinician might adopt a 'medical model' approach – that is, by reducing outcomes to narrow biological markers. The person undergoing treatment, however, may have other ideas. This is particularly apparent when it comes to long-term conditions. In these cases, 'success' can become ambiguous. Unable to be 'cured', subjective value is often instead invested in more abstract, perhaps distant outcomes, and in the potential trade-offs between, for example, potential longevity, quality of life, and flexibility of lifestyle. In short, the environment of chronic disease forces us to ask what medicine is actually *for*.[8]

Take, for example, a case study described in physician Jay Katz's 1984 book *The Silent World of Doctor and Patient*. Iphigenia (not, of course, her real name) is a newly engaged 21-year-old. She has recently been diagnosed with breast cancer. She is informed that she must undergo a full mastectomy, which is considered the 'correct' treatment. She hesitantly agrees, but is approached by her doctor the night before the operation, full of misgivings about the prospect of someone so young being forced to undergo such a life-changing procedure. The doctor goes on to outline several potential alternative therapies that she is unaware of, but which would not require the removal of her breast. After some consideration, Iphigenia decides to cancel the surgery and instead chooses the much less invasive option of a lumpectomy alongside radiation therapy – she will remain outwardly unscarred but might perhaps live with a greater chance of the cancer returning.

Later addressing a panel of physicians and associated professionals, Iphigenia reaffirms her decision and expresses joy at being able to begin married life without disfigurement. Despite this, her account is met with hostility by those present, many of whom are incredulous that she has been allowed, with no medical training, to decide upon a therapy perceived as inferior.[9] Katz does not record Iphigenia's eventual fate, because in truth it does not matter. His point is that she was, in a sense, lucky. A temporary breakdown of the paternalistic framework had allowed her to meet her own subjective needs in a way that she would never have been able to had things proceeded according to the familiar script.

Iphigenia's case was unusual for the time. For much of the twentieth century, those undergoing medical care were understood to occupy a passive 'sick role'. Perhaps most famously articulated by American sociologist Talcott Parsons in 1951, this implied 'withdrawal into a dependent relation'.[10] In exchange for treatment, so the thinking went, the individual was expected to subordinate themselves fully to the authority of the professional, leaving their own perspective at the door. Even where two doctors might disagree, Parsons argued, 'the layman is not qualified to choose between them'.[11]

As Katz makes clear, however, this approach fails completely to take into account the subjectivity of healthcare. Each individual has their own precise definition of what is meant by 'successful'. Someone diagnosed with an almost-certainly terminal cancer may, for example, choose to tolerate the side-effects of gruelling chemotherapy in an effort to pro-long their lives. Conversely, another person with the same diagnosis may reject treatment, accept their likely fate, and decide to live as well as possible – however they define it – for the time they have remaining. Neither decision is 'wrong', but for most of the twentieth century only one would have been approved of by mainstream medicine.

So how does this relate to insulin and the management of diabetes? Insulin therapy is always a complex undertaking. There are few simple, isolated decisions to make in the management of diabetes. Those affected must attempt to integrate their treatment with the demands of their everyday lives and their own subjective needs, but it is rarely possible to achieve this seamlessly. As a result, they must engage in an almost permanent process of value judgement and refinement as they decide which sacrifices should be made and when.

As we have seen throughout this book, people with diabetes cannot be dictated to. They are rarely the bedridden, caricatured 'sick' envis-aged by figures like Parsons. There is no way, in practice, for any doctor to enforce obedience. Their treatment becomes a self-administered part of their everyday lives, and they can choose to conduct it however they like.

Insulin renders this power even more potent. The choice of when and how much of it to inject is integral to how it interacts with the body, and to the outcomes of treatment. Users are, as a result, able to manipulate this to whatever end they desire, regardless of professional opinion. Insulin, as a result, forces subjectivity to centre stage, revealing as it does the messy reality of life with diabetes.

Almost nobody using it sticks rigidly to the rules and guidelines set out by healthcare professionals at all times, even if, ideally, they would prefer to. This has been a feature of insulin therapy since its inception. It is certainly not the niche realm of individuals like Jack Eastwood, who, as described in Chapter 3, sought to fundamentally engage with their treatment as, essentially, lay-scientists. More often, it is the product of subjective factors. Monica Winn, who was diagnosed as an 8-year-old in 1927, for example, remembers varying her food and insulin as her lifestyle demanded from at least the early 1940s.[12]

It is worth taking the time here to examine some case studies in more depth. As a young mother with T1DM in the 1980s, for example, Gillian Clifton regularly allowed her blood sugar to become higher than her doctors recommended – deliberately eating slightly more food or taking slightly less insulin to ensure that this would happen. She did, however, have good reason:

> My first marriage broke down when my daughter was very young, and I worked full time. I would be up at the crack of dawn to get her to nursery. Cycle to work, work all day, cycle home, pick her up, do all the stuff with my daughter then have stuff to do for work. I couldn't – I didn't have time to go hypo. I wasn't comfortable going hypo when I was on my own in the house with her. So my blood sugars were running higher than they should have been to avoid that happening.[13]

In Clifton's case, her decision was based upon a crucial subjective value judgement. Alone with her daughter and aware that a severe episode of hypoglycaemia could threaten them both, she made a concerted effort to prevent this from happening regardless of the consequences for her blood sugar. Implicitly, her daughter's health was an integral part of her own.

Similarly, Vic Marriott was diagnosed with T1DM in 1955 at the age of nine. A working-class man, he changed careers regularly throughout his life, and it was not always easy to reconcile the expectations of managing his condition with the demands of his often laborious work. He frequently used to deliberately 'run a bit high' while in physically demanding employment to avert the risk of hypoglycaemia – for example in a demolition job that involved climbing high up the exterior of buildings.[14] While his doctors recommended against this kind of employment, Marriott was in no position to refuse it, and could expect little flexibility on the part of his bosses. Like Clifton with her daughter, he was forced to prioritize. In a position where an episode of hypoglycaemia could very well lead to a deadly fall, he weighed up his options and acted accordingly. The similarities here are unmistakable – Marriott responded to the demands of a job that he relied upon in much the same way that Clifton responded to the demands of motherhood. Insulin therapy does not exist apart from the broader social reality of the person undergoing it, but reflexively interacts with it.

This remains true even for those in comparatively privileged positions. Frank Kaye, for example, lived a very different life to Marriott when he was diagnosed with diabetes in the mid-1960s at the comparatively old age of 27. Kaye was the son of a Glasgow businessman, and he worked in his father's furniture shop for his entire life before eventually inheriting it. Unsurprisingly, he did not face the same pressures that his subordinates might have. As a senior member of the company, and later its owner, he was answerable only to his father and to himself, while the specific duties of his job were not particularly physically demanding.

Given this context, Kaye's continued livelihood was almost certainly never in any real jeopardy should his condition cause problems at work – by requiring, for example, a few extra breaks, or causing occasional tardiness. Unlike his working-class contemporaries, he was often able to much more easily structure his day around the demands of his treatment.

Nonetheless, this was not always possible. Kaye was also expected to live up to the sometimes hyper-masculine social and professional expectations of a 1960s businessman. He frequently entertained clients, which usually involved taking them out to restaurants where he occasionally struggled to keep to his prescribed diet. He was also often expected to drink heavily – something that he preferred not to do, feeling that it did little good for his blood sugar levels.

In the end, Kaye found a clever workaround to this. He began to take aside one of the serving staff whenever he ate out, explaining the situation and asking that, the next time one of his party insisted upon another round of whisky, he be brought a small measure of ginger beer in a tumbler instead, leaving no one the wiser. This attempt to reconcile his diabetes management and his professional life was successful, and, amusingly, may actually have increased his standing amongst clients – from their perspective it looked like he could hold an incredible amount of liquor with few ill-effects.[15]

These subjective decisions are not always made with particularly large stakes. They are also important when it comes to seemingly mundane but personally meaningful quality-of-life issues. Marriott sometimes maintained a higher-than-recommended blood sugar level when working, but he also did so during his free time when the situation warranted it, and made a point of not telling his doctors – implicitly because he knew they would not be pleased:

> If I knew I was going to go all-night fishing I made sure I wasn't going to go hypo. So I'd reduce insulin to make sure. I would rather be high than go hypo. On the end of Newhaven Pier I am in the middle of the bloody night, you know! . . . I don't think [the doctors] knew about it actually . . . and I seemed to get away with it pretty well.[16]

In the 1970s, Carol Cowan made a similar decision to ignore her strict lunchtime meal prescription because doing so opened up the possibility of spending the money on something far more personally desirable:

> Well, in sixth year at school my mother gave me dinner money . . . which I did not [use to] pay for dinners. I used to pinch two digestive biscuits and an apple out of the house every morning and that was my lunch, and I kept the dinner money for two Carlsberg Specials at the dancing on the Saturday![17]

Cowan's attitude was little different to many British teenagers, and no doubt her friends did similar. As someone with diabetes, however, her rebelliousness was always implicitly medicalized, and had she been caught she would almost certainly have faced accusations of negligence both for skipping her lunches and for going out drinking. However, the ability to enjoy a night out with her friends clearly meant a lot to her. Being able to do so was one important metric by which she measured the 'success' of her overall management. To Cowan, the quality of life she gained by going dancing outweighed what she lost by disobedience.

The process of entering adulthood and beginning to live independently can highlight the subjectivities associated with insulin therapy particularly starkly. One place where this is particularly apparent is amongst those who have recently left home to attend university – a transition that can be difficult for anyone, regardless of their health. Students undergoing insulin therapy are subject to considerable additional pressures due to their condition. Like their peers, they are thrust into an unfamiliar environment to which they are expected to acclimatize quickly, often shorn of the support networks that may have characterized their prior lives. As a result, they are often keen to quickly establish a place for themselves – to become a part of their new community. This, however, comes with a certain vocabulary and grammar. When they get to university, new arrivals – atomized as they are – often attempt to cultivate a collective identity by engaging in behaviour that they broadly agree is 'typical' of student life. In the UK, for example, this stereotype can often involve pronounced hedonism in social life – particularly when it comes to habitual binge drinking. By engaging in similar activities, new students effectively construct and maintain their identity via their bodies.[18]

In a 2009 article, sociologist Myles Balfe argues that such 'body projects' have particular relevance to students with T1DM.[19] There is, he suggests, a relatively consistent anxiety amongst this group that they will be perceived as somehow not 'normal' – that their condition will be read as incompatible with the 'typical' student lifestyle, and will as a result prevent them from integrating successfully. In order to mitigate this, Balfe claims, they often make extensive use of 'body projects' to project a valorized, 'healthy' image of themselves in which diabetes, while it exists, plays little role.

One of Balfe's interviewees, for example, articulated this succinctly, claiming that when new people witness him injecting insulin on a night out, 'they can't believe I would go out and drink so much and do the things that I do and have diabetes as well'.[20] These kinds of 'body project' however, can have obvious consequences for maintaining stable blood glucose, but they do not imply any crude denialism. Most of Balfe's interviewees were acutely conscious of their condition and its potential long-term repercussions. Their choices were the product of a complex subjective thought process that sought, however imperfectly, to resolve these contradictions in a way that was most acceptable to them.

Almost everyone engages in a 'body project' of one form or another to cultivate, maintain, and project a desired identity.[21] One of the most overt and visible examples of this can be found in the bodybuilding community. Unlikely though it may seem, this subculture also illustrates the subjectivities involved with insulin use – this time, however, outside of the scope of diabetes.

That some athletes use performance-enhancing drugs is a widely acknowledged and controversial phenomenon. Less well known, however, is that insulin is sometimes one of them.[22] Decorated Serbian bodybuilder Miloš Šarčev, for example, is open about his use of the substance, and claims to have injected it as part of his training regimen continually from 1993 to 2003. Šarčev believes that this posed little risk. It is clear that he has done a lot of research on the subject, and he approaches insulin in a systematic way. In short, his thinking is that, when exercised, muscles require considerably more oxygen than usual. As a result, during particularly intense activity, blood flow to them is many times what it would be at rest. This, he suggests, provides an opportunity. If the blood is saturated with nutrients alongside 'a potent agent that's going to shove them into every . . . cell' during this period, highly efficient growth can be promoted.[23]

Unsurprisingly, Šarčev is a controversial character, and he admits that medical professionals are often highly critical of his approach. On this, he is absolutely correct. Many doctors consider the use of insulin in this fashion incredibly dangerous.[24] The greatest threat, most of them warn, is serious hypoglycaemia. Taking medically unnecessary doses of insulin on top of what is already being made by the pancreas practically invites blood sugar levels to fall precipitously, particularly when combined with intense exercise. There have, in fact, been several tragic bodybuilder deaths in which this phenomenon appeared to have played a central role.[25]

For his part, Šarčev does not dispute that insulin *can* be deadly. While he believes that it can be a safe and effective tool, he also accepts that using it comes with risks. He insists, however, that the danger can be minimized. The problem is not, he claims, the substance itself, but user error – difficulties emerge only when people fail to respect it, incorporating it into their training without being fully aware of how it interacts with their bodies:

> You don't know how to do it – if you don't know how to drive a car . . . you're going to crash probably. So it is not insulin that is dangerous . . . I only use insulin when it makes sense, when you want to shove all the nutrients . . . into muscle tissue.[26]

Šarčev has little time for those who attack his approach. Whatever his critics say, be they other bodybuilders, doctors, or scientists, he maintains that the strategy can be highly effective. The topic remains hotly debated within the bodybuilding community, and representatives from both the pro- and anti-insulin camps frequently lock horns on the issue, but there is no doubt that the practice is now widespread.[27]

While it used to be a relatively niche interest, the use of insulin by bodybuilders is becoming increasingly visible in the mainstream. Today, even a quick Google search returns numerous resources outlining guidelines for those interested in trying it, alongside countless YouTube videos and posts on specialist forums.[28] One *British Journal of Sports Medicine* article from as far back as 2003 even estimated that a full quarter of bodybuilders who used performance-enhancing drugs of one kind or another also used insulin.[29]

There are some fascinating parallels between insulin use in the bodybuilding and diabetes communities. As both testimony from figures like Šarčev and the many guides to 'protocols' available online demonstrate, a significant proportion of insulin-using bodybuilders do not take the practice lightly. Instead, they often engage in it very carefully indeed. In their 2005 book *Dr. Golem*, Harry Collins and Trevor Pinch discuss the use of performance-enhancing drugs amongst bodybuilders. Many of these athletes, they argue, are far from ignorant. Rather, they draw from an extensive body of 'remarkably detailed folk pharmacological knowledge' and, even where their interpretation of the literature diverges from orthodox opinion, they are often able to confidently argue their position.[30]

Whether these bodybuilders are 'right' or 'wrong' is beside the point here.[31] It is certainly true that, used inappropriately, insulin can be considerably more immediately dangerous than other performance-enhancing drugs, but this is the case whoever takes it. While it is not always easy, countless people with diabetes manage to survive perfectly well using insulin every day, particularly once they have become sufficiently knowledgeable about its properties. Sometimes, as we have seen, the lay expertise that they develop throughout their lives leads them to adopt patterns of use that contradict professional opinion. On occasion, these approaches come, in time, to be vindicated. Are bodybuilders – or at least the ones who *do* make a concerted effort to study and understand insulin – altogether so different?

Medical professionals often find themselves exasperated when faced with articulate insulin-using bodybuilders. The dangers seem clear: even the most enthusiastic figures acknowledge that the practice can never be *perfectly* safe. So why take the risk at all? In short, the answer is usually that those risks seem acceptable when weighed up against the potential benefits. Just as someone with diabetes may manage their condition in a way that clinicians feel is less than ideal for broader subjective reasons, the bodybuilder may too – accepting the potential danger as a fair price for effectively cultivating the muscular physicality so integral to their identity.

While it is used in some other niche environments, insulin will always be most relevant to people with diabetes. The complex way in which it interacts with the body, and the ambiguity of terms like 'success' and 'failure' in treatment lend it a powerful, subjectivity-enabling character. This sets it apart from most pharmaceutical products.

People with diabetes are often called upon to make important decisions about how to implement therapy, and in doing so must manage a variety of competing demands, their physical health and the expectations of medical professionals (which may or may not be the same thing) being only two relevant factors in this process. This can be stressful, but it also allows them to meaningfully shape treatment according to their own individual needs in a way that is not possible in the context of many other conditions both short and long term.

But what, exactly, does this have to do with DAFNE and the widespread transition to 'patient-led' treatment at the beginning of the twenty-first century? In order to appreciate the importance of lay-subjectivity in this development, it is necessary to consider the broad culture of medical practice in the years leading up to it.

Paternalism

In the late 1980s, figures like Berger – who actively sought to challenge the traditional authority of physicians over all aspects of treatment – remained relatively isolated internationally. Despite the influence of discourse about SBGM and limited 'self-adjustment', the majority of doctors in the British Isles, where DAFNE would appear only a decade later, continued to educate their patients in an often piecemeal fashion that generally reinforced the traditional, paternalistic relationship.

Berger's ideological position had not, however, emerged in a vacuum. It was directly influenced by liberatory, anti-authoritarian social movements with their origins in the counter-culture of the 1960s and 1970s. Medicine was only one target of this political groundswell, but it felt the development keenly. By the 1980s, a vibrant literature had emerged to engage critically with concepts of power and subjectivity in healthcare.[32] Crucially, many of these contributions came from self-reflective medical professionals. Katz, discussed in the previous section, falls into this category, for example. So too does Arthur Kleinman, now a highly influential figure in the field of medical anthropology, whose prolific efforts to explore the social, cultural, and philosophical worlds of patient experience betray a deep suspicion of traditional, paternalistic attitudes.[33]

Nonetheless, for many practising physicians this kind of writing seemed a relatively niche, academic interest. For some, however, it proved highly influential. In Britain, SBGM pioneer Robert Tattersall, for example, found much to admire in the discourse it prompted, and he took the time to consider its implications for diabetes management. In 1989, he, with his Czech colleague Michal Anděl, wrote a scathing critique of contemporary practice in the journal *Diabetic Medicine*. The brief paper is damning, and it is perhaps little surprise that only a few years later a disillusioned Tattersall chose to retire.[34] It contained, however, a pointed reflection on the implications of paternalism for diabetology:

> Whether they have insulin-dependent or non-insulin-dependent diabetes, not everyone is able or willing to follow the strict conditions for achieving good metabolic control. When people wish to opt out of the grand scheme, they are described as 'noncompliant' for which the most usual treatment is, as in a totalitarian society, re-education which may even include internment for 'stabilization'. There are in fact many reasons why people do not strive for perfect diabetic control ... in these patients the benefits of good blood glucose control may be counterbalanced by increased hypoglycaemia, neuroticism, anxiety and what the patient sees as an intolerable interference with social, sexual or professional life. To rebel against the grand scheme requires considerable courage and usually results in the rebel being stigmatized as a dissident or non-complier.[35]

Tattersall and Anděl acknowledged that 'success' in diabetes management is often subjective, ill-defined, and shaped by a multitude of factors beyond the strictly 'medical'. Equally important, however, they implicitly suggested that physicians – by adopting an overly paternalistic attitude – could actually do great harm. In a subsequent issue, Tattersall clarified this position while responding to a piece of correspondence from Glasgow endocrinologist Eon Hamish McLaren:

> I agree that one reason for rejecting a doctor's advice may be robust independence but another, when one is seen by relays of different doctors, may be pure self-preservation. It has been said that non-compliance is the best defence against iatrogenic [medically inflicted] disease and confusion.[36]

By alienating their patients through paternalistic micromanagement, he thought, doctors only encouraged them to take their advice with a pinch of salt. If they dismissed whatever was said in the clinic and got on as they always had, people with diabetes could effectively shield themselves from the stress, anxiety, and sometimes the shame that might result from trying to earnestly, or perhaps critically, engage with authority figures uninterested in their subjective needs or insight.

Margaret Howie, for example, remembers doing just this in her mid-twenties, at around the same time Tattersall and Anděl's article was published. Howie, who had lived with diabetes since the age of nine, knew by experience that the advice her doctor was giving her would not work. Instead of arguing, however, she politely listened and agreed despite having no intention of following through:

> They'll say things like 'oh, you're a wee bit high in the morning, so you need to increase your long-acting at night'. But what they didn't realize was you'd kind of already tried that, and it didn't work, because then you were hypo in the morning instead of slightly high. So it was a question of kind of nodding your head and saying 'oh yeah, I'll fiddle about with it and try it', knowing that it wasn't going to work and then going off and basically just carrying on the way you were.[37]

This kind of behaviour almost certainly occurred throughout the history of insulin therapy. Prior to the 1980s, however, few would have admitted it. In 1977, for example, 'N.D.' wrote in to *Balance* to explain how, at 71 years old, thirty-three of them with diabetes, they had adopted a relatively laissez-faire approach to treatment with little in the way of routine, occasionally altering 'food intake or insulin as the [urine test] indicates'. Despite stating that 'no one knows as much about my diabetes as I do myself', however, 'N.D.' admitted to having 'momentary qualms', particularly when reading about the strict regimens used by others. 'Am I right', the letter ends, 'to avoid rigidity and continue a method which I believe brings reasonable control?'[38] It feels almost as if 'N.D.' is asking the editors of *Balance* to grant approval. Acting in a way that contradicted mainstream medical opinion – and in doing so openly undermining their presumed authority – was rare, and always invited sharp criticism.

In the 1980s, however, far more examples begin to emerge of individual laypeople openly engaging with and criticizing, if not always their own doctors, the paternalistic character of the medical profession. Originating in the same social movements that had inspired Berger, the idea of 'patient rights' had now gained considerable traction, creating space for the emergence of, as historian Alex Mold puts it, 'the individual, autonomous patient', able to 'articulate new demands about their ability to determine how they lived, and about the fate of their bodies'.[39]

The consequences of this can be seen clearly in the response to a 1987 *Balance* article by sociologist David Kelleher, which asked 'why people with diabetes don't always follow their doctor's instructions.'[40] The piece prompted a furious response in the following issue from Alexandra Weston, a woman with T1DM, whose letter barely attempts to hide its contempt for Kelleher's entire premise:

> I hope that all my BDA subscription does not go on projects such as that of David Kelleher! To begin with, he has started on completely the wrong track by researching into why people with diabetes do not 'obey' their doctors' 'instructions' . . . Mr Kelleher seems to think that diabetics should be thinking all the time about their diabetes and arranging their lives around it, whereas personally I do very much the opposite, and I very much doubt if I am the only one! I always value greatly my doctor's help and advice, but like to have the freedom to adjust my levels as I see fit, after eight years' experience.[41]

Weston is perhaps a little defensive here. Kelleher certainly did appreciate the limitations of simplistic 'medical model' thinking, and acknowledged the relevance of subjectivity. He made a point of highlighting, for example, that 'some diabetics have different ideas about things from their doctor, and are more inclined to follow their own ideas . . . [and] in some cases they have different beliefs about health and illness'. Nonetheless, the fact that this was clearly an emotive topic is telling. Weston clearly had no patience for what she perceived as reductive, authoritarian paternalism.[42]

While the critical engagement that came with such burgeoning assertiveness was no doubt valuable, the elephant in the room had gone nowhere. By the 1990s, it seemed increasingly likely that, whatever their relationship with their doctors and their approach to treatment, most people with diabetes in Britain were at significant risk of long-term complications and early death.

As professionals began to recognize this, it had palpable consequences for the way they wrote about diabetes. Patient guidebooks from the 1990s often omit the detailed exchange lists that had so characterized earlier publications. While they do tend to suggest that meal plans should be determined with the help of a dietitian, they usually forgo the idea that food should be precisely weighed and measured. In her 1992 book *Diabetes: A Beyond Basics Guide*, physician Rowan Hillson explained the reasoning behind this:

> The problem is that our bodies are not machines. A car owner can calculate the number of miles his car goes per gallon and knows how much the fuel tank holds – so he knows how much fuel he needs and how often. But there are so many variables in the working of the human body that a simplistic view of food as a fuel may lead to a false sense of security.[43]

Consequently, she thought there was 'little point in weighing out precise carbohydrate portions for every meal.' There is a certain resignation to Hillson's tone here.[44] Strictly prescribed diets, she thought, were by definition flawed because insulin needs never remained the same day after day. But this also implied that a high vulnerability to long-term complications might be, in practice, inevitable for most.

One strategy that might, some thought, have at least a chance of mitigating the risk, however marginally, was to encourage 'healthy eating' in a very general sense. If people were vulnerable either way, then attempting to minimize the non-diabetes related risk factors made sense. As Stephanie Amiel, a former member of the DAFNE Study Group, puts it:

> We'd gone . . . to a much more laissez faire attitude of 'let's just teach them healthy eating, because they're a high cardiovascular risk'. And for several [years] – at least a decade – patients weren't taught anything except the principles of healthy eating, which told them nothing about how to use the insulin with regard to the food![45]

Hillson encouraged her readers not to be frightened of their condition and downplayed the risk of complications, but this served primarily to maintain morale.[46] Behind the scenes, the mood was more melancholy. Overall mortality had, it could be said, fallen since the mid-twentieth century. This, however, was largely down to innovations in treating and slowing complications when they did occur. On the whole, diabetes in the 1990s continued to be understood as a 'disease with a poor prognosis'.[47]

This did prompt action. In 1989, for example, clinicians, government representatives, and people with diabetes from across Europe came together in Italy for a meeting hosted by the World Health Organization (WHO) and the IDF. The result was the St Vincent Declaration, a broad agreement that established several collective goals. The declaration called upon individual countries to invest in programmes for the detection and treatment of diabetes and its complications, raise public and professional awareness about the condition, support research, and ensure adequate education for those affected.[48] Despite its noble intent, it quickly became apparent that achieving these objectives would be no easy task, particularly as many governments were reluctant to provide the additional funding necessary. 'Sadly', Harry Keen recounted in 2000, 'by the time we were due to turn recommendations into implementation, the process began to fray.'[49]

That same year, the UK government-commissioned review *Testing Times* made it clear that there was still a long way to go. In England and Wales at least, it found major shortcomings in provision, alienation between healthcare professionals and their patients, and far fewer individuals being given adequate education than was considered satisfactory.[50]

In a 1995 article, however, Berger and Ingrid Mühlhauser once again highlighted the positive results that they were seeing with the Düsseldorf model, pointing out that one of its great strengths was that it could be adapted to almost any economic circumstances. One of their colleagues, E.G. Starostina, had, for example, managed to successfully implement a version of the programme in Moscow in 1994, right in the midst of Russia's post-Soviet economic collapse.[51] 'Especially in the light of the St Vincent Declaration', they argued, 'we should be optimistically striving forward to implement intensified treatment programmes for Type 1 diabetes in all . . . countries once they are of proven efficacy and safety.'[52]

This paper had appeared in the BDA-affiliated journal *Diabetic Medicine* – a publication most British diabetologists tried to remain up to date with. The article prompted interest, and for good reason. The German approach could potentially kill two birds with one stone. First and most obviously, it provided a framework by which 'intensification' might be achieved in practice. Second, it worked with, not against, the contemporary trend towards increased lay-assertiveness and attention towards subjectivity.

When DAFNE did launch publicly in 2002, freedom from paternalistic restriction was, in fact, central to its marketing – encapsulated in the tagline 'eat what you like, like what you eat'.[53] Perhaps unsurprisingly, this was one feature highlighted with concern by sceptics. Amiel, for example, recalls a meeting with the Department of Health in which one government representative incredulously insisted that it was irresponsible to tell people with diabetes to eat whatever they like.[54] Simon Heller, her former colleague in the DAFNE Study Group, also remembers some stern opposition from physicians:

> We had loads of pushback! I mean, I remember going up to Scotland and speaking to certain consultants and them saying 'what is all the fuss about? This is just rubbish'. And it was immensely frustrating . . . It took so much longer that we would have hoped, but people began to realize that this was the only approach which could work.[55]

DAFNE's implementation in Britain marked the beginning of a dramatic shift in mainstream approaches to diabetes management.[56] Not everyone, of course, attended a course. Even today, only a fraction of those with T1DM do. Nonetheless, it is now common practice to encourage newly diagnosed people to make use of similarly autonomous treatment regimens utilizing carbohydrate counting and individual adjustment whether or not they enlist in one of the formal DTTPs.[57]

When I was first diagnosed with T1DM in 2009, for example, I was encouraged from the outset to adapt my treatment according to diet and lifestyle – no one ever attempted to impose a prescribed diet. I was even given a DAFNE pocketbook, which contained estimated carbohydrate values for a wide selection of different food types to assist in dose adjustment.

The introduction of 'patient-led' treatment did not transform British diabetology overnight. In fact, it faced considerable opposition, as Amiel and Heller highlight. Today, however, it is considered the benchmark for insulin therapy, and much energy is now put into discussing ways in which those who remain on more traditional regimens might be persuaded to adopt it.[58]

DAFNE, like the Düsseldorf model that inspired it, fundamentally reworked the roles and responsibilities of both medical practitioners and their patients, turning the traditional relationship on its head and recasting laypeople as the driving force in treatment. In doing so, this demanded that professionals accept a more distant role – reconceptualized as remote sources of support with little involvement in day-to-day treatment decisions. Given the historically authoritarian character of diabetology in Britain, how can we explain this apparently uncharacteristic about-face?

There is, perhaps, a rather stark irony here. For Berger, the Düsseldorf model could not be decoupled from his broadly left-wing political stance. Despite this, the implementation of similar approaches in the UK owed much to the increasingly insidious ideological hegemony of neoliberal capitalism, in the context of which they needed imply no meaningful rejection of paternalism at all.

Neoliberalism

When Conservative leader Margaret Thatcher won the 1979 UK general election and was installed as prime minister, she promised a radical new approach to the nation's economy. Ostensibly committed to individual freedom, autonomy, and choice, her neoliberal approach to governance saw her embark upon a crusade of privatization in the service of capitalist class interests, slashing public spending and mortally wounding the trade union movement as she did so – mirroring Ronald Reagan's government in the United States.

What, precisely, is meant by neoliberalism as an ideology, however? Scholars continue to add nuance and granularity to the precise definition of the term.[59] Nonetheless, almost all agree with some variation of Stephanie Lee Mudge's argument that uncompromising marketplace logic is its primary intellectual feature, supported by a political commitment to 'liberalization, deregulation, privatization, depolitici-zation, and monetarism'.[60] In short, it seeks – in theory – to curtail government involvement in economic matters, with a view towards promoting free market competition. In doing so, however, it also attempts to frame itself as essentially apolitical – as Thatcher herself put it, 'there is no alternative'.[61]

To this end, the neoliberal project goes well beyond economics. It also attempts to reshape the way we understand ourselves and our position within society. Undermining collectivist thinking in favour of a stark individualism in which everyone is, first and foremost, a consumer, it encourages us to interpret every aspect of human life according to marketplace logic. David Harvey perhaps articulates this best in his influential 2005 book *A Brief History of Neoliberalism*, sum-marizing it as:

> A theory of political economic practices that proposes that human well-being can best be advanced by liberating individual entre-preneurial freedoms and skills within an institutional framework characterized by strong private property rights, free markets, and free trade ... entailing [also] much 'creative destruction' ... not only of prior institutional frameworks and powers ... but also of divisions of labour, social relations, welfare provisions, technological mixes, ways of life and thought, reproductive activities, attachments to the land and habits of the heart ... hold[ing] that the social good will be maximized by maximizing the reach and frequency of market transac-tions, and ... [seeking] to bring all human action into the domain of the marketplace.[62]

By the last decade of the twentieth century, and particularly fol-lowing the end of the Cold War, neoliberalism had come to permeate Britain's institutions of state. The NHS, which, from 1948, provided universal healthcare to all British citizens, remained free at the point of use, but it too had been transformed by the new political consensus. 'Patients' could no longer be understood simply as beneficiaries of care, but rather as a distinct category of consumer. The notion of the 'patient-consumer' did not by any means originate during the 1980s, but a particular, marketized version of the concept proved highly useful to Thatcher and her successors, who cultivated the idea that healthcare should be understood as just another product. 'Technocratic approach[es] to health service delivery', as Mold puts it, were replaced with 'business methods and market mechanisms'.[63] As ever, this was often couched in the language of individual liberty. The *Patient's Charter*, a consultative paper produced in 1991 by the early John Major government, for example, firmly laid out the 'rights' of those who used the healthcare system, attacking the traditionally paternalistic culture of the NHS and promoting the idea of a new approach to medicine based, ostensibly at least, on principles of individual agency and 'choice'.[64]

When New Labour came to power in 1997, it did not meaningfully challenge the contemporary ideological consensus. Before becoming prime minister, Tony Blair had abandoned the party's commitment to far-reaching public ownership, and throughout his leadership positioned himself as uncompromisingly pro-market.[65] While the new government did provide some additional funding to the NHS, it maintained an essentially neoliberal trajectory.[66]

The new government quickly discovered that, when it came to health, complex, long-term conditions of all stripes now constituted a significant burden on the health service. The cost of treating T2DM and its complications alone, for example, cost an estimated £2 billion per year – a full 4.7% of the total budget.[67] Responding to this, it commissioned a task force in 1999, directing it to investigate potential new policy directions to meet the challenges of the new millennium. Two years later it published its conclusions. *The Expert Patient: A New Approach to Chronic Disease Management for the 21st Century* recommended that, so far as was possible, individuals should be encouraged to take an active role in their own care via extensive education provision:

[Those] with chronic diseases need not be mere recipients of care. They can become key decision-makers in the treatment process. By ensuring that knowledge of their condition is developed to a point where they are empowered to take some responsibility for its management and work in partnership with their health and social care providers, patients can be given greater control over their lives.

Self-management programmes can be specifically designed to reduce the severity of symptoms and improve confidence, resourcefulness and self-efficacy.[68]

This position aligned with neoliberal individualism, and, perhaps more importantly, it also implied reduced costs – the authors pointing out that 'a number of studies suggest that user-led self-management programmes lead to a considerable reduction in visits to General Practitioners and other health professionals'.[69]

From 2002, the government began to implement these recommendations with the Expert Patients Programme (EPP) – an education initiative for those with long-term health conditions.[70] EPP classes, which were delivered by laypeople, sought to create confident 'graduates' who would require less supervision and would be able to communicate their questions, concerns, and needs more articulately to healthcare practitioners when the need arose.

The creation of such conceptual individuals was, however, a controversial one, and was opposed by medical professionals across the board. In one 2004 editorial for the *BMJ* tellingly named '"Expert patient" – dream or nightmare?', Joanne Shaw and Mary Baker highlighted the scale of resistance, citing a private survey in which only 21% of responding doctors thought the proposals wise. Contrary to the government's task force, many professionals worried that the 'expert patient' might in fact cost more to treat and give them more work to do, while doing little to improve their practice. Shaw and Baker provide some indication as to the root of this anxiety. To many doctors, it seemed like the proposals would only encourage the most demanding and unreasonable of their patients:

We know from reading the press and listening to the debate that when doctors come across the term 'expert patient' they hear different things. For the chief medical officer, expert patients are 'people who have the confidence, skills, information and knowledge to play a central role in the management of life with chronic diseases.' The suspicion is that for many doctors, the expert patient of the imagination is the one clutching a sheaf of printouts from the internet, demanding a particular treatment that is unproved, manifestly unsuitable, astronomically expensive, or all three. Or, possibly worst of all, a treatment the doctor has never heard of, let alone personally prescribed.[71]

Reading between the lines, it is obvious that, to some, the principle of 'self-management' appeared to pose a threat to professional authority. Shaw and Baker attempted to resolve this anxiety, but their conclusion only emphasizes the endemic paternalism that continued to exist within the British medical profession: 'Long live expert patients – but, in the interests of doctor–patient relations, let us find something else to call them.'[72]

Unsurprisingly, the EPP was often implemented in a half-hearted way, and observers pointed out that, for all its lofty talk of centring laypeople, and 'despite governmental enthusiasm', 'often professionals cling to power in their engagements with patients, controlling information and dismissing [their] efforts . . . to theorise or explain their condition.'[73] Some had even argued to this effect when it was still in the planning phase. Patricia Wilson, for example, believed that despite its formal commitment to 'the rights and responsibilities of those with chronic illness', the programme had 'no corresponding strategy to challenge professionals' assumptions towards [them]'.[74]

Despite the influence of neoliberal ideology, twenty-first century medical professionals clearly exhibited many of the same paternalistic attitudes that they always had, and they were deeply suspicious of so-called 'expert patients'. How, then, can we explain the overwhelming success and influence of 'patient-led' management in the context of diabetes?

DAFNE emerged out of same milieu as the EPP, and it was even featured as a case study in *The Expert Patient* – used as an example of what lay participation could achieve in practice. However, while it did receive some funding from the more general programme and embraced much of the same rhetoric, it was never formally part of it.

One of the reasons for this was that DAFNE, in short, attempted to create a very specific *kind* of 'expert patient'. In contrast to the EPP, which tended to focus on relatively subjective peer support and promoted lifestyle adaptations to lessen the impact of living with long-term conditions, DAFNE utilized a structured training programme, delivered wholly by health professionals, with the stated objective of equipping participants with distinct insulin and diabetes-related skills that would allow them to manipulate their treatment effectively.

Despite the centrality of flexibility in DAFNE's public relations, many of those involved with establishing the programme in the UK continued to think primarily in 'medical model' terms. Quality-of-life benefits were welcome, of course, but, implicitly, they were of secondary importance. Amiel, for example, somewhat uncomfortably admits that, while she 'would not wish to divorce flexibility from part of the package', she 'would not put it above getting a good medical outcome' because, as a doctor, the latter is what she believes she is employed to do.[75]

That much of the public-facing literature discussing DAFNE made a point of highlighting first and foremost the potential freedom that it offered made sense in the context of neoliberalism. It was, to use the language of the ideology, a clever piece of marketing designed to pique the interest of potential consumers. By framing the programme as not only a path to improved long-term clinical outcomes, but also one way of achieving more immediately tangible benefits, it worked to justify and mitigate the additional labour demanded of those who adopted it.[76]

This is not at all to suggest that any of the people responsible for establishing DAFNE in the UK were committed ideologues for either neoliberalism or reductionist 'medical model' approaches to healthcare. While he agrees with Amiel that, as a doctor, he must always attempt 'to help people help themselves to prevent long-term complications', Heller, for his part, is infectiously enthusiastic about the 'horizontal, non-hierarchical' way in which he is able to interact with 'expert patients' in his practice.[77] Clearly, these were doctors trying to do their job as well and as thoughtfully as possible, albeit in ways that were shaped by the broader context in which they operated.

There is no doubt that DAFNE was a very welcome development for many. It is not difficult, for example, to find positive feedback online, some of which describes the course in superlative terms.[78] Some people, however, were more critical. As recently as 2020, one disappointed former attendee claimed that they found it difficult to translate the training they were given into practice.[79] Another, in 2014, reflected that they were simply underwhelmed. While they felt that it would probably be valuable for those who 'really had been left to fend for themselves', and expressed shock that some of their coursemates did not appear to know 'what a carb was', they did not feel that it offered them any new insight.[80]

When he began to introduce the concept of DAFNE to the general public, Heller had quickly learned that 'around ten percent of people had been doing this already and [had] taught themselves', but had often kept quiet, usually to avoid a lecture from their doctors.[81] For these more confident people with substantial personal expertise, the course could be frustrating. What about it prompted these feelings?

In Düsseldorf, Berger had become aware very early that some people were unilaterally taking their treatment in radical new directions. 'Long before diabetologists had begun to debate the liberalization of the lifestyle of type 1 diabetes patients', he later reflected, 'our patients had come to these conclusions and were attempting to put them into practice.'[82]

Berger rejected the notion of any 'rigid rules' in the context of insulin therapy. He sought to create a cooperative relationship in which lay expertise would complement professional knowledge. In this sense the DTTP was only the beginning. While it did provide specific training, its implicit long-term purpose was to encourage its attendees to become highly confident at self-management, providing a foundation upon which they could build as they, with their doctor, worked ever further to refine their technique.

This way of looking at things makes less sense in the neoliberal context of DAFNE, where it is cast, predictably, as a product. As a 'skills-based education programme', one article in 2016 argued, for example, it offered regional NHS commissioning authorities the opportunity to 'achieve cost savings of £93,133 per 100,000 population annually' while allowing them to fulfil National Institute of Clinical Excellence (NICE) guidance that stated all people with T1DM should be offered access to structured training.[83] This ostensibly depoliticized, technocratic framing was simply not compatible with Berger's grand liberatory project. Such marketized logic is the key to understanding some attendee dissatisfaction with DAFNE. As Sara Glasgow has argued, neoliberal approaches to health delivery, particularly in the context of 'patient-led' long-term care, are often marked by a distinct sense of contractualism:

Participants in this regard are brought to view and manage their life-style in such a way as to mimic the relations between agents in a market environment – the fulfilment of obligation for the generative end; in this case, not of wealth but rather of health and well-being.[84]

In this framework, outcomes are understood to be contingent upon the fulfilment of certain obligations. Sometimes, this is rather explicit. Take, for example, CSII treatment. Since 2002, this has been available through the NHS – though not always easily.[85] For those who have been able to acquire a pump in this manner, however, it has often been subject to many terms and conditions. One 'agreement' form, still freely downloadable from the website of the Association of British Clinical Diabetologists, for example, outlines strict expectations of recipients. They are asked to accept an initial six-month 'loan' period and to acknowledge that subsequent provision will be conditional on them meeting 'the agreed targets for improvement in blood glucose control and/or reduction in hypoglycaemia.' Furthermore, if they 'do not benefit from the pump, or persistently fail to use the pump safely and effectively, the use of the pump may be reviewed and funding for the pump consumables may cease and [they] may be required to return the pump to the Trust.' There is space for the new user to sign and date the document, and for a witness – a staff member – to do the same.[86]

This is not an isolated example. It is now quite common for people with diabetes to use continuous glucose monitoring (CGM) technologies. These are powerful SBGM devices that are attached to the body, providing constant blood glucose feedback. They are able to record daily fluctuations with considerable accuracy, and often allow insulin doses and food intake to be logged in the system as well. They can, as a result, be very useful for identifying blood sugar patterns and adapting treatment accordingly.[87] They are, however, expensive. In the UK, the Freestyle Libre – one of the cheaper options – costs around £50 per sensor if bought privately, and this will last around two weeks. Potentially, then, this represents costs of approximately £1,200 annually. Recently, however, this device has become available on the NHS.[88] Once again, however, acquiring it is not always as straightforward as asking.

In 2021, one Twitter user shared a photograph of a document that they were expected to sign to access Libre in Birmingham, which stipulated several obligations for potential users. They were, for example, expected to use the device at least eight times per day and attend regular review sessions. If they failed to do so, or if their subsequent results were not 'satisfactory', there was always the threat that it might be taken away. All the while, the idea that recipients were being given significant responsibility was highlighted in no uncertain terms:

> Understand that further funding is dependent on the results achieved with the Libre2® device which will be reviewed at 3–4 and 6 months post initiation. YOUR DIABETES IS NOW YOUR RESPONSIBILITY MORE THAN EVER. You will therefore need to demonstrate adequate use of this system and ask for help if needed.

Like the CSII 'agreement', there is space for a signature and countersignature. In this case, the document overtly acknowledges its transactional nature, asking the reader to declare (by signing) that they have 'understood the information above and agree to this contract prior to starting/continuing on Freestyle Libre® through NHS funding'.[89]

This contractualism is often palpable even where it is not made so explicit. A few months after my own diagnosis with T1DM in 2009, I lost the DAFNE-branded pocketbook that I had been given. When I called the organization to ask for a replacement, they were highly reluctant – even when I offered to pay. This was frustrating, because I had found the carbohydrate tables very useful! While I did finally persuade them to send another in the post, what arrived was slightly different to the one I had been given initially. Its contents page indicated that certain pages contained guidelines about dose adjustment, obviously taught during the actual DAFNE course. These sections, however, had been torn out!

At the time, this decision seemed rather strange, but, reading DAFNE in a neoliberal light, it makes a great deal more sense. I had not attended a course. As a result, I had not agreed, implicitly or explicitly, to any obligations. While the original document had contained only carbohydrate tables, this new one incorporated advice derived from the taught aspects of the course – material I, apparently, had no right to.

Similarly, while doing the research for this book, I thought that it might be useful to examine a copy of the handbook used during the DAFNE course for myself. However, I found it almost impossible to find one. They are simply not available to those who have not attended for themselves. Little wonder there are numerous comments online about how keenly the group protects its 'secret knowledge'![90]

As a product in the neoliberal marketplace, DAFNE must be sold as a discrete commodity – a DTTP that uses a defined set of rules and guidelines to produce certain outcomes for consumers, be they people with diabetes or regional NHS authorities. This leaves it unable to effectively deal with lay expertise that contradicts its teaching. As one former attendee in 2014 complained, the thing that really annoyed him was 'the refusal to accept, or even discuss any alternative approaches and to consider that they might be equally effective or better'.[91]

One 2014 Irish study, for example, interviewed several DAFNE 'graduates'. One young woman explained how she continued to struggle despite implementing the rules that she had been given:

> No matter how much I follow all the ... points of DAFNE it still didn't help, but it was just that I was at a loss, you know, my blood sugars were still very high readings, you know, and there was no answer for the high readings, you know? ... So I found it frustrating.[92]

DAFNE provides training geared towards flexibility, but it does so in a relatively inflexible manner. Its vision does not extend, as Berger's did, to encouraging people with diabetes to go *beyond* the rules that they are taught – becoming truly autonomous and self-sustaining by developing their own insight and rejecting those guidelines that do not work for them. People are, instead, told how they should approach their condition and expected to adapt their treatment accordingly.

While many people have found DAFNE extremely useful, it is easy to see why some confident and experienced participants become frustrated where their lay expertise contradicts what is being taught, particularly where the latter is presented as objectively 'correct'.[93] Looked at in this sense, DAFNE bears a perhaps unlikely resemblance to R.D. Lawrence's idea of the 'diabetic creed' – both attempt to define, codify, and enforce particular ideals of treatment enforced by soft power.

In this sense, the authoritarian attitudes that had characterized much of twentieth century diabetology have not gone anywhere – they have simply been relocated from individual prescription to a packaged body of educational principles that people with diabetes are expected to follow as they conduct their own care. According to the contractual logic of neoliberalism, this also implies considerable responsibility. If they fail to meet their obligations – by not abiding by the defined terms – then who else but they can be held responsible for any deterioration that might subsequently occur? The more things change, the more they stay the same.

In twenty-first century Britain, however, people with diabetes *are* often in a position to dynamically engage more than ever before with their treatment according to personal expertise and subjective need. The distant and essentially toothless authorities that have always defined diabetology have become increasingly so, and they now rarely attempt to exert direct control over their patients' daily lives, while the average person using insulin has vastly more powerful tools with which to assess, evaluate, and refine their treatment.

Meanwhile, while the educational principles they utilize are not particularly flexible, DTTPs like DAFNE *do* provide those who attend with a considerable amount of subject-specific knowledge and encourage thoughtful engagement with treatment. For those who previously struggled with elements of their management, this training at least has the potential, in some cases, to serve as a first step to more radical and confident engagement – whether or not in ways approved by professionals.

Nonetheless, the subjective decisions people with diabetes make are not necessarily respected where they do not align with 'medical model' thinking. Professionals continue to exert moral authority when it comes to the definition of value in management.[94] While many *do* now make a serious effort to engage with individual subjectivities and with the broader value systems of their patients, there are limits. Even medical ethicists 'committed to respecting patient subjectivity', philosopher Amélie Oksenberg Rorty points out, 'typically also offer specific normative and regulative principles to guide "rational choice" in medical contexts'.[95]

In Britain, neoliberalism has had considerable influence on health-care at both the organizational and cultural level. Nonetheless, this has occurred in the context of the NHS which, for now, continues to ensure that basic supplies are available free to those who need them. People living in countries that ensure medical provision for their citizens often take it for granted, but it is an immense privilege. Across the planet, many – even most – cannot rely upon their governments to cover the cost. In some cases, this is the result of political instability, economic deprivation, or both. In others it is the product of deeply embedded free-market ideology. Regardless, it can have devastating material consequences for those who require insulin.

Chapter 5

The Insulin Crisis, 2002–Present

In 2017, almost a hundred years after insulin was 'defensively' patented for the express purpose of preventing unethical profiteering, Alec Smith was found dead in his Minneapolis home at only 26 years old.[1] He had been diagnosed with T1DM two years previously, but seemed to have adapted well to the demands of insulin treatment. What had gone wrong? Smith's great misfortune was that he had been lucky enough to be born a citizen of Earth's wealthiest nation. Unlike most industrialized countries with the finances to do so, the United States offers no universal health coverage to its population. When he ran out of insulin and could afford no more, he was simply left to die.

This tragedy was far from unforeseeable. While a small number of Americans are able to access subsidized healthcare through government programmes such as Medicare and Medicaid, almost all working-age people rely on a fully privatized, insurance-based marketplace to meet their medical needs.[2] Those fortunate enough to have stable, full-time employment often receive health insurance as a perk of the job, but this leaves many to fall through the cracks.

When he was first diagnosed in 2015, Smith was registered with the insurance that his mother held through work. According to the terms of the 2010 Affordable Care Act (ACA), this provided him with coverage until his twenty-sixth birthday. Unfortunately, he was already approaching the cut-off date. In 2017, he 'aged out', and was left to fend for himself.[3]

Smith did not live in poverty. He worked as a manager at an independent restaurant, and received a salary of around $35,000 per year: not an extravagant income by any means, but comfortable enough for a single man. Losing his health insurance, however, was devastating. As a small business, his employer offered no coverage, but he was paid too much to qualify for Medicaid assistance. His only options were to either take out an individual health insurance policy, or to go without and pay for any costs up front.

It quickly became obvious that for Smith to maintain the level of care he had enjoyed on his mother's insurance, even the most 'affordable' policy would involve crippling costs. In addition to a $450 per month premium, he would be expected to pay full price for his supplies until he met an annual deductible of $7,600. Even after all of that, he would still be liable for smaller 'co-pay' fees every time he picked up a prescription until he met his 'out of pocket maximum'.[4] In short, Smith was looking at a potential yearly bill of at least $13,000 – over a third of his total income.

It must have seemed absurd. For only $5,400 per year, he could buy the right to spend up to another $7,600 on medical care alone, and his other expenses – rent, utilities, food, etc. – had not gone anywhere. At work he had heard talk that the owners planned to open branches at several new locations, and, as the ACA had also stipulated that companies with more than fifty employees should offer health coverage to at least 95% of their full-time workforce or face penalty fines, he reasoned that, as a manager, he would almost certainly be offered insurance when the move went through.

After his diagnosis, Smith remembered being shocked at an initial pharmacy bill of around $500 for a month's worth of insulin and supplies. This was a lot, but it must have seemed manageable. It was, after all, only $50 more than he would be paying in premiums if he took out an individual policy. His mother later remembered him telling her how he planned to go without insurance, asking 'how bad could it be?'

The answer, as it turned out, was very. Insulin is now one of the single most expensive substances on the planet. When he went to the pharmacy to pick up his prescription, he was told that a single month of supplies would set him back $1,300 – quite a sum for a substance that one recent article suggested could be produced – profitably (!) – at an average cost of only $133 per user per year.[5]

Smith was shocked. He had, with some justification, never imagined just how expensive such an apparently ubiquitous thing might have become. He simply did not have the money that was being asked of him and, lacking insurance, was not eligible for any of the commercial discount schemes that might have reduced his immediate costs, however modestly. In the end, he left with only a fraction of what he needed to effectively manage his condition.

With little other option he began to ration what little he could afford, injecting as little as he felt he could get away with while also radically cutting his carbohydrate intake. If he could buy enough time to reach his next pay packet, he must have thought, at least he could replenish his supplies and take stock of the situation. This decision, however, proved to be disastrous. Smith's strategy was a desperate gamble, and, as it turned out, a fatal one. Only a month after being unceremoniously dumped from his mother's insurance policy, he fell into a coma and died of DKA. Damningly, his experience was far from unique.

When Allen Hood, for example, turned eighteen, he was rejected by Medicaid. In the end, he also turned to insulin rationing. Two years later his mother came home to find him on the floor, unconscious. Cruelly, she was forced to watch for over two hours, only days before Mother's Day, as paramedics tried in vain to save her son.[6]

Jada Baldwin was also uninsured. When she was taken to hospital with DKA in 2019, she admitted that she had been unable to give herself insulin for three whole weeks (!) prior to her admission. She was sent home, and died eight days later.[7]

After he lost his job in November 2017, Jesse Lutgen's health insurance went with it. The most 'affordable' individual plan he could find at short notice came with an incredible $10,000 deductible so, like Smith, he tried to pay out of pocket. He was found dead in February the following year.[8]

These stories represent only a tiny fraction of those who have died for want of a vial that, twenty-five years ago, could be bought for less than the cost of a round of drinks at most mid-range bars.[9] At the time of writing, typing 'insulin' into the fundraising website GoFundMe returns countless results, many of them last-ditch attempts to acquire vital supplies from uninsured (and underinsured) Americans. Make no mistake, this is a crisis.

It is not, however, by any means a uniquely American one. Global diabetes rates are increasing rapidly, and demand for insulin along with them. As the planet's richest country, however, the United States makes for a telling case study. The argument that the resources or infrastructure to provide universal health coverage is simply not there holds little water. If the political will to do so existed, the authorities in Washington D.C. could end the crisis for their constituents almost overnight several times over.

Moralism and Prejudice

Unlike universal healthcare more generally, demands for insulin accessibility can be muddied by the common perception that diabetes is somehow *not like* other health conditions – that those who develop it, and those who die as a result, are not just unfortunate, but have a direct hand in their own suffering. While incorrect, understanding the root of this position and the political discourse of which it is part is essential to making sense of what is happening.

The earliest accounts of diabetes are rarely particularly moralistic. From around the seventeenth century, however, this begins to change. By the nineteenth, rapidly fatal cases – that is, cases of T1DM – had become the exception, not the rule, and diabetes as a whole came to be strongly associated with elderly, large, and often rich people. Physicians changed their tone accordingly, often laying blame at the feet of an ever-more indulgent society that had, they thought, long forgotten the wisdom of moderation and humility. As historian Arleen Marcia Tuchman describes in her 2020 book *Diabetes: A History of Race and Disease*, in the late Victorian period it did not take long for this to intersect with contemporary ideas about race.[10] White people – and particularly Jews – were considered most susceptible, while many felt that Black and Indigenous populations were essentially immune.[11]

The problematic arguments used to explain this were typical of the period. Non-Whites, for example, were often understood to be protected because of their 'primitive' qualities.[12] This highly racialized perspective echoed – probably not coincidentally – claims that had long been made about the labouring classes more generally.[13] Even after Oskar Minkowski and Josef von Mering convincingly implicated the pancreas as the 'seat' of diabetes in the body in 1889, some physicians – such as US neurologist George Miller Beard – continued to argue that nervous disruption could play a major role in causing it.[14]

Figures like Beard believed that those who occupied the upper rungs of the social hierarchy – at the time almost universally White – were highly vulnerable to nervous strain because of their status as 'brain workers'. 'On the highly civilized man', he wrote in 1881, 'there rests at all times a three-fold burden – the past, the present, and the future!'[15] This supposed weight, so his thinking went, could lead to systemic disruption – potentially resulting in diabetes among other conditions. By contrast, uneducated 'muscle workers' engaged in manual labour were considered insulated from risk because, implicitly, they lived a relatively simple, ignorant life in which they were rarely required to think. Attitudes to non-Whites, in the context of contemporary race science, reflected a similar – albeit essentialized – perspective. Supposedly less intelligent, they were simply not considered capable of overstraining their nervous material to any meaningful extent. 'The utter want of curiosity in matters that do not come immediately home to them', Beard wrote of Indigenous Americans, for example, 'is a feature in their character most noticeable and most interesting, contrasting, as it does with the excess of Yankee curiosity.'[16] Similar, he thought, were the 'Indians of South America and Central America – the negroes of Africa and of our own country, young children everywhere, and adults who have never matured in the higher ranges of intellect'.[17]

Associating nervousness and diabetes served an important ideological function. By casting the condition as a product not, necessarily, of gluttony and overconsumption, but, perhaps also of thinking, working, and caring *too much*, it could be reconceptualized as evidence of a noble constitution.[18]

This is not to say that concerns about overconsumption and reckless indulgence had disappeared by the late nineteenth century – 'nervousness', for those who agreed with figures like Beard, was simply another potential risk factor, not the only one. In 1890, Charles Wesley Purdy felt perfectly comfortable attributing rising death rates from diabetes to immoderation, writing that, after the American Civil War, 'the people entered upon a career of luxurious living', which 'largely accounts for the decided impetus given to the disease during the period named'.[19]

By the late nineteenth century, Jews were considered by far the group most vulnerable to diabetes. In part, this was attributed to the widely held contemporary stereotype that they were a particularly 'nervous' people.[20] However, many also felt that their lifestyles put them at additional risk. In 1898, for example, one 'Dr. Bond', responding to a conference paper given by Elliott Joslin and his colleague Reginald Fitz, suggested that 'the Hebrew races probably have this disease more than any other people on account of their love of high living . . . They are given to parties, they congregate together and have frequent and irregular meals.'[21]

While its association with 'nervousness' waned from the 1920s, the image of diabetes as a condition of the affluent White remained. This, once again, was often attributed to their apparently extravagant, sedentary lifestyles. In 1924, for example, Haven Emerson claimed that:

Diabetes is essentially a disease of the idle rich, in which definition idleness and riches are not opposites to occupation and poverty but terms applicable to anyone whose environment and self-support does not require him some sustained bodily exertion during the day, and whose earnings or income permit him, and whose inclination tempts him, to eat more regularly than he needs.[22]

This narrative did not meaningfully change for several decades. From the 1950s and 1960s, however, research involving Indigenous communities such as the Akimel O'odham – often referred to by contemporaries as the Pima, the name given to them by early Spanish colonists – seemed to indicate that they experienced particularly high diabetes prevalence rates, contradicting prior assumptions that they were rarely at risk.[23]

This was eagerly seized upon by the emerging field of human genetics, whose representatives quickly – though not always particularly convincingly – proposed that a particular inherited gene might be responsible. Tellingly, it did not take long for this to be framed in the language of 'progress' and 'civilization'.

Indigenous people – American or otherwise – were, contemporary geneticists suggested, particularly vulnerable because they often possessed a so-called 'thrifty gene'.[24] Supposedly preserved more readily in members of less 'advanced' societies that had not had time to adapt to industrialized living, this supposedly ensured that they were able to efficiently store energy as body fat – an important protection against potential disruption to the food supply. However, where someone with this gene was transplanted – as many Indigenous people had been – into an urbanized environment, so the thinking went, it came to represent a problem. Despite a steady food supply and few if any periods of scarcity, their body would continue to store energy as if it were preparing for an imminent famine. Obesity, and with it diabetes, threatened as a result.[25]

By the 1980s, the 'thrifty gene' hypothesis had gained significant traction, and it had become rare to associate diabetes primarily with well-to-do White people. Instead, a firm link was increasingly being drawn between it and more marginalized non-White communities. In 1985, the US government-commissioned Heckler Report formally acknowledged that ethnic minorities were at a disproportionately higher risk than White Americans, experiencing a 50% higher overall mortality rate. Those from Indigenous communities in particular, the report claimed, were in fact a staggering ten times more likely to develop the condition.[26]

Of course, in reality non-White populations had always been affected by diabetes to one extent or another. Studies alluding to this cropped up time and again throughout the early twentieth century. Joslin, for example, commented on its prevalence amongst Indigenous Americans in 1940.[27] Statistician Louis Dublin noted as early as 1928 that, contrary to mainstream opinion, Black people – and particularly Black women – were developing diabetes in extremely high numbers.[28] As Tuchman points out, however, this research tended to fall on deaf ears.

Ideological factors have always played as much of a role as any supposedly 'neutral' science in shaping the context of diabetes research and management. Discourse around it reflects, as a result, contemporary prejudices and class antagonisms in a rather clear way. Both working-class and non-White populations were largely invisible because medicine itself was a predominantly White, upper middle-class profession governed by, and in the interests of, the White upper middle-class.

For much of the early twentieth century, diabetes received considerable interest *because* it seemed, at first glance, to so keenly threaten that very demographic. Would it have been the subject of so much research had it been considered primarily a condition of the marginalized? Would insulin itself have been isolated when it was? It is impossible to know, but these are questions worth asking.

Whatever the true historical prevalence rates, it is now beyond question that the poor, and particularly the non-White poor, are disproportionately affected. In the USA, 37.3 million Americans live with diabetes of one kind or another, with the prevalence rate amongst the White population around 7.5%. By contrast, the figures for Indigenous, Black, and Hispanic communities sit at 14.5%, 12.1%, and 11.8% respectively.[29] On the other side of the Atlantic, the picture is similar. In 2009, Diabetes UK reported that the most deprived members of British society are two and a half times more likely to be diagnosed than average, with those from Afro-Caribbean and South Asian backgrounds at substantially higher risk.[30]

Are certain ethnic groups inherently at higher risk of developing diabetes because of their genetics? Perhaps. But is that all there is to it? This was certainly the impression given by the Heckler Report, and later public health messaging adopted similar framing. In Tuchman's concluding chapter, for example, she highlights one ADA poster from the 1990s. 'Diabetes', it reads, 'favours minorities'. They should, it suggests, therefore discuss ways of mitigating the threat with a doctor.[31] The message seems to be that they simply *are*, objectively, at higher risk, and that there is little to be done about it beyond encouraging individuals to be considerate of their health. As Tuchman is quick to highlight, however, the poster's racialized framing, along with its implicitly individualistic message, fails to acknowledge the importance of socio-economic context. This is a serious blind spot. While it is now clear that some non-White people are more vulnerable, the idea that genetics alone can be held responsible for this represents a dangerously superficial reading of the situation.

Over the last hundred years, the vast majority of cases of diabetes have been T2DM. Today, it represents a full 90% of all diagnoses. While there does appear to be a genetic component to this, it is now widely accepted that diet and lifestyle can also play a role, and that obesity specifically is a particularly significant risk factor. This goes some way towards explaining why diabetes once seemed so prevalent amongst the affluent. They did not want for food, could live relatively sedentary lives, and, by comparison to the poor, often had very rich diets.

Today, however, the wealthy are well aware of these risks, and, crucially, they have the time and money to make use of this information. They can eat high-quality, nutritionally balanced meals, commit to regular exercise regimens, and attend frequent medical screenings. T2DM is not, of course, entirely a product of lifestyle any more than it is of genetics, and lots of wealthy people will develop it regardless, especially as they age. The demographic as a whole, however, is able to mitigate the threat, should they so choose, and can rest easy knowing that, should it become necessary, they have access to high-quality medical care.

The poor, however, have no such luxury. Where more prosperous families might employ others to do their household cleaning, look after their children, and even cook their meals, this is not an option for those on more modest incomes. When they get home from their jobs – and today, many low-paying occupations are nonetheless relatively sedentary – there is still much work to be done: housework, childcare, etc. Time is often at a premium, and what little of it there is offers limited opportunities for an exhausted worker who has only more of the same to look forward to the next day.

Compounding this, even those who would *like* to spend their free time preparing healthy meals and exercising can find cost a major barrier. Gym memberships and sporting equipment can be extremely expensive, and, while fresh produce might appear superficially cost effective, it often perishes quickly and sometimes requires relatively advanced cooking skills to make most effective use of – skills that the most overworked may not have the time or resources to cultivate.

Processed ready-meals and takeaway food are often proportionately very cheap, fast, and filling, and in many cases they are undeniably tasty. It may seem absurd on the face of it, but when one commentator called the McDonald's McDouble burger 'the cheapest, most nutritious and bountiful food that has ever existed in human history', the claim was probably not too far from the truth. Costing only a single dollar when it was introduced in 1997, it provided almost four hundred calories.[32]

Shifts in the class context of both food and working culture now expose the poor to multitudes of risk factors for diabetes that their wealthier peers are able to easily sidestep. The rich may still engage in conspicuous consumption using cuisine, but they are now much more likely to show off by enjoying complex dishes created by technically accomplished chefs than with sheer quantity and richness.

The prevalence of diabetes amongst poor and often non-White populations must, therefore, be understood, in large part, as a product of this class context. It may be true that some people – and perhaps some ethnic groups – are more genetically vulnerable than others, but the deck is stacked against them by poverty and discrimination from the beginning.

With this in mind, the continued popularity of concepts of greed and overindulgence in diabetes discourse – despite clear evidence that it is now vastly more common amongst the underprivileged – seems particularly troubling. If poor people do not want to get diabetes, so the thinking seems to go, they should simply eat healthily and exercise more. The demographic shifts in prevalence are now widely recognized, but nevertheless the implicit link to lazy gluttony remains – albeit repackaged along inverted class lines. The poor, implicitly, are cast as feckless simpletons, unable to act in their own best interests.

This, of course, is absurd 'bootstraps' thinking. The most marginalized populations are not able to simply wave a magic wand to remove themselves from the hardships of their economic position. Nonetheless, this perspective has direct relevance to the insulin crisis. According to its troubling framing, if diabetes can be prevented by moderation, then the problem must be largely manufactured. With adequate self-discipline, the suggestion seems to be, insulin itself would be rendered unnecessary in most cases, and the issue would resolve itself.

It is important to point out here that there continues to be considerable public confusion about different 'types' of diabetes, their causes, and the way in which they are treated. As by far the most common form, however, T2DM has shaped the popular narrative to a much greater extent than T1DM or any other variety. Predictably, some have attempted to create a moral dichotomy here. There are plenty of examples of people with T1DM expressing frustration at those with T2DM for the moralism they are perceived as bringing to discourse around diabetes as a whole, many of whom insist that the different 'types' of the condition should be given more distinct names to prevent any confusion.[33]

In a 2014 study investigating the experience of stigma amongst those with T1DM, for example, one interviewee openly admits that she is 'not a big fan' of those with T2DM. 'I've got no time for them really', she says, 'because . . . I've tried my hardest and I've got something that I've got no say in and then there's millions of dollars spent on people that could have prevented it.' A man quoted in the same article agrees, claiming that 'there's the fat lazy type and there's the type that I've got', the former of which he often describes in 'slightly derogatory terms'.[34]

The overall tone of these complaints echoes Chris Morris' satirical separation of HIV/AIDS into 'good' and 'bad' versions – the former acquired via blood transfusion, the latter from sexual activity or intravenous drug use – in 1997's *Brass Eye*. Those with T1DM are cast as innocent victims, contrasted against people with T2DM, who have only themselves to blame. As Morris implicitly highlights, however, this is a less than helpful position. Perhaps there is an argument to make for a firmer distinction between T1DM and T2DM in both popular and professional discourse – the two are after all distinct conditions with their own challenges, and it is important that people are aware of this. Nonetheless, no one 'chooses' to get diabetes of any stripe. Given the socio-economic (and often racialized) context of T2DM, many arguments of this kind achieve little more than to compound the burden of an often-already marginalized community. Understanding and solidarity, not resentment, is of paramount importance here for all parties.

In any case, we can at least say that T2DM *is* often related, to some extent, to dietary and lifestyle factors, whatever pressures are behind them. T1DM, on the other hand, is not. Nothing (yet) in our power can prevent the autoimmune destruction of the pancreatic islet cells or the resulting need for lifelong insulin therapy.

As the crisis has received increasing public attention, however, the moralistic attitude that often colours discourse about diabetes as a whole has produced some spectacularly short-sighted arguments concerning T1DM. Some commentators, almost certainly imagining that they are being very helpful, have even suggested that in order to reduce medication requirements, they should also simply moderate their life-styles and eat less.

If this seems like something of a throwback to the early twentieth century, that is because it is precisely that – sometimes explicitly. In 2021, American physician David Ludwig, for example, took to Twitter to share an article of his that self-consciously sought inspiration in the carbohydrate-restricted diets of the Frederick Allen era. 'Today,' he argued, 'technology, not diet, dominates diabetes management . . . [yet it] continues to extract a huge toll on patients and public health, with economic costs in the US nearly $1 billion a day.'[35] Referring to another piece that he had previously co-authored for the *Journal of Clinical Investigation*, Ludwig suggested that 'to improve health outcomes and reduce costs, the future of diabetes lies in a centuries-old approach: a low-carbohydrate diet'.[36] In short, while he does not suggest outright starvation, he does advocate a return to the kind of highly restricted dietary regimens used before, and immediately after, the introduction of insulin. Where early twentieth century physicians (incorrectly) believed that this was necessary because excess carbohydrate caused deterioration by 'stressing' the islet cells, Ludwig promoted the theory, in part, as a response to the accessibility crisis. While most people who reacted to his Twitter post were highly critical, a small minority expressed support. Kal Chinyere, an Atlanta physician, for example, enthusiastically approved of the overall message, asking 'if we cut their insulin requirement in half don't we cut their costs in half?'[37]

The premise that carbohydrate restriction might reduce insulin requirements is not incorrect. Eating substantially less will, in most cases, do just that. This, however, is beside the point. The amount of insulin prescribed by physicians increased consistently in the early twentieth century because it allowed people to eat enough to satisfy, rather than merely sustain, themselves, while also allowing them to abandon the unpalatable high-fat diets that had been common in the 1920s. In calling for a return to strict carbohydrate restriction, Ludwig appears also to suggest the reassertion of a drearily paternalistic approach to management centred on obedience to inflexible dietary prescription.

The almost unanimously negative response to Ludwig's article from those actually using insulin was unsurprising because his argument missed the point in an exceptionally clumsy way. In the midst of widespread calls for accessible insulin and anger at the political reality that thousands of people were (and are) being mercilessly exploited – sometimes to the point of disability and death – by its manufacturers, he had chosen to seize the opportunity to show solidarity by promising salvation via an exciting mixture of pious self-deprivation and overbearing, morally loaded authoritarianism, while implicitly embracing the neoliberal argument that the problem could be solved through 'better' individual choices.

Insulin therapy is, in practice, a complex undertaking, but its core principle is quite simple – it is simply about replacing (or supplementing) a hormone. A high quality of life and substantial freedom to eat and live as desired is not mutually exclusive with maintaining stable blood sugar levels, and there is no inherent moral value in needing more or less insulin. Enough of it, used appropriately, makes both possible. Blame for the crisis cannot be individualized, but must fall upon the organizations whose policies ensure that some lack the healthcare they require, and the political powers that enable them.

Power and Profit

Whatever the 'type', diabetes is a dangerous condition that can lead to disabling and even potentially fatal long-term complications. In reality, however, it need not – with enough insulin it can be effectively controlled. People living with diabetes do, on average, live slightly shorter lives than the general population. Nonetheless, their life expectancy is now substantially greater than it was in the mid-twentieth century, and there is no reason to believe that this gap might not close further in the future, should we allow it to.[38]

Over the last hundred years, diabetes should have lost much of its bite. With effective treatment it is an unpleasant inconvenience and often a chore to deal with, but it should be little more than that. However, for many it remains an existential burden, and it continues to be recognized as one of the leading causes of death worldwide. This framing, however, is somewhat misleading. When people die after rationing insulin, 'diabetes' is usually listed as the cause. When they develop avoidable complications that end up killing them, the same thing happens. It is true that in most of these cases the condition has physically contributed to their deaths, but many would have survived had they been able to access the supplies that they needed. Often, they did not die 'of diabetes', but rather of inadequate access to healthcare. The problem here is political, not biological.

Those who argue against universal – or at least more affordable – healthcare usually do so out of an ostensible commitment to free market economics. On paper, the free market is one in which private companies can set whatever price they choose for their products, while individuals are able to freely decide where to buy what they need. In theory, this leads to a community of rational consumers who continuously seek out the best deals, encouraging businesses to undercut one another to attract customers. This, so the thinking goes, leads to affordable prices and greater efficiency.

This utopian, self-regulating free market is, of course, nonsense, and insulin serves as a perfect example of why. Those who use it, and especially those with T1DM, are a captive audience. They cannot choose to go without. Furthermore, global production is dominated by only three major companies – Eli Lilly, Novo Nordisk, and Sanofi.[39] In practice, these manufacturers hold a near-monopoly, but they rarely undercut one another as the theory suggests they should. Instead, they do the opposite, forming an effective cartel that is able to maintain artificially high prices. If their customers have no choice but to buy their products, why ruin the party by competing when they can work together to guarantee one another's profits?

There are some ways in which insulin can be acquired more affordably. Doctors can pass on 'samples' to their patients, for example, which of course also function to advertise certain branded products. Sometimes 'free clinics' can help in the short term, and pharmacy coupons or the so-called 'co-pay' cards issued by manufacturers might also reduce immediate costs. None of these, however, are real solutions, and they tend to come with a long list of terms and conditions. 'Co-pay' cards, for example, are often not available to the uninsured, and even where they are, the holes in the system become rapidly apparent.[40]

During the COVID-19 pandemic, for example, Eli Lilly announced that it would make a card capping monthly insulin 'co-pays' at $35 available to those without insurance. Should someone acquire one of these cards and have it accepted by a pharmacy, neither of which is a given, they might still be expected to pay hundreds or even thousands of dollars per year for their supplies. There is also a $7,500 annual limit, based on Lilly's 'contributions' towards the still-extortionate list price with each purchase. In practice, this means that virtually nobody is able to get everything they need at the reduced price.[41]

'Co-pay' cards are also never issued simply out of goodwill. They also function as effective marketing schemes that encourage people to buy certain products – and still at a significantly inflated cost. They may save money for individuals in the short term, but manufacturers more than recoup their costs in the process, and never fail to cynically use the opportunity to considerable PR effect by highlighting how much they apparently care.[42] This is not the only way companies work to repackage profit-oriented policies as charitable benevolence.

Eli Lilly, Sanofi, and Novo Nordisk each, on paper, donate vast quantities of insulin to those in need through what are known as patient assistance programmes (PAPs). Each company has its own with slightly different small print, but they operate similarly in practice and provide one avenue through which some uninsured people who meet certain criteria can receive their prescriptions free of charge, apparently direct from the manufacturer. This sounds very generous on the face of it, and undoubtedly some people have benefited from insulin acquired through PAPs. We should, however, be very careful about attributing any social consciousness to profit-seeking enterprises. No private insulin manufacturer makes its products for the good of humanity, but rather to secure the greatest possible return on its investment – that has been clear since the 1920s. PAPs are no exception. In fact, they represent a very clever piece of misdirection indeed.

When someone receives insulin from, say, Lilly Cares – Eli Lilly's PAP – they are not actually getting it from Eli Lilly the manufacturer. Instead, it comes from the Lilly Cares Foundation, an affiliated, but ostensibly independent, non-profit organization. Eli Lilly proper provides thousands of dollars' worth of insulin to Lilly Cares, which then distributes it to successful applicants. Why the middleman? The answer should surprise no one. When Eli Lilly 'donates' insulin to Lilly Cares, they are able to designate this as a charitable contribution. This 'donation' can then be claimed back as tax relief at 'fair market value' – a value which is determined entirely by the manufacturer cartel of which Eli Lilly is a controlling member!

Insulin may be sold for far more than it costs to produce, but its 'value' here is determined wholly by its list price. As prices go up, these generously minded companies can claim back ever more from the government, even while their expenses remain stable. The system is framed as manufacturers doing people a favour, but in reality the function of PAPs is clear. This is a clever system of tax avoidance dressed up as humanitarianism.[43]

Even if we ignore the motivation behind them, and focus specifically on the material help that they can sometimes provide, neither PAPs nor coupons nor any other manufacturer-sponsored scheme offer meaningful solutions to the insulin crisis. They are inefficient, flawed, and unsustainable. The often-Byzantine eligibility requirements differ between manufacturers and providers, and they often do nothing to provide the additional equipment – SBGM devices, for example – that many rely upon. Most important, however, is that they operate only for as long as the companies that run them want them to. People reliant on their products are left at the mercy of the free market and, as we have seen, this mercy extends only so far as it remains compatible with corporate profit. If they decided that it was in their interest, pharmaceutical companies could tear up their various assistance schemes in an instant – and their track record suggests that they would do so without hesitation.

But how do the manufacturers defend the astonishing list price of insulin? Can they, somehow, justify the extortion? Asked to explain their high costs, the go-to response almost always includes some variation of the claim that it is necessary to enable innovation: that the expense is the only way to ensure that new technologies capable of further improving diabetes management can be developed.[44] Some free-market fundamentalists even build on this by suggesting that the government itself is the problem, suggesting that the cost of meeting regulatory requirements both increases prices and stifles the development of new technologies.[45]

It does not take much sustained analysis to show that none of this is true. In fact, the two arguments are mutually contradictory. If government approval *is* so costly that it discourages innovation, then why exactly are high prices required at all? If there are going to be high prices anyway, then where is the innovation? In reality, there have been precious few meaningful breakthroughs since the early 2000s, and what exists now is perfectly sufficient for the vast majority of people using insulin. The important thing is getting it into their hands.

Over the last twenty-five years, the only real technological developments made by the major manufacturers have been the development of new, more expensive versions of insulin. In many of these cases, they were often responsible for little of the scientific work involved, instead using their considerable financial reserves to purchase the rights to technology developed by other companies.

Humulin, for example, was the first biosynthetic 'human' insulin designed to resemble that made by the pancreas. First developed in 1978, from 1983 it was sold as an alternative to animal-derived versions by Eli Lilly. In reality, however, the company had played little part in its development. Most of the actual innovation had been made by Genentech, another biotechnology firm. Lilly's only major role had been having the foresight to recognize potential profit, and the wealth to ensure that it could buy the licence.[46]

In 1996, Humalog signalled the introduction of so-called analogues to the market. These newer varieties, which now account for most of what is used in wealthier countries, are tweaked to have much longer or shorter windows of action and intensity than older versions. Of course, they also have much commercial value. These newer, lab-made varieties are completely different to older formulas and to one another, allowing them to be patented in their own right.

One area where manufacturers *have*, in a sense, innovated, is in making sure that their insulin remains profitable. By making small, largely insignificant 'improvements' to the original formula – a process known as 'evergreening' – they are able to protect their long-term financial interests by artificially extending patents well past the usual twenty-year limit, even though the products they cover remain practically indistinguishable.[47]

Biosynthetic insulin and analogues have been particularly profitable for the major manufacturers because they have been very successful in marketing them as a direct upgrade to older versions. When Humulin came onto the market, for example, it did not take Eli Lilly long to announce that animal-derived varieties had become obsolete and subsequently withdraw them from sale, leaving the more expensive synthetic formula the only option.[48] This perception of obsolescence has also contributed to a widespread lack of enthusiasm amongst companies that might produce generic versions of out-of-patent formulas.

Analogues were met with much initial scepticism from physicians. A few studies conducted in the mid-2000s did suggest that they might modestly reduce HbA1c levels by comparison to standard biosynthetic 'human' varieties, but the evidence was far from unequivocal.[49] What, then, was the point? To many, these expensive new 'designer' insulins seemed to represent yet another cash-grab by the pharmaceutical industry. There was almost certainly some truth in this suspicion. Analogues certainly did offer greater financial rewards for the manufacturers, so a certain amount of healthy cynicism is definitely appropriate. In this case, however, those critical physicians demonstrated their own ideological limitations in a largely unhelpful way.

While the measurable differences between analogues and older varieties of insulin appeared modest, the metrics used in studies comparing them focused almost exclusively on outcomes as defined by the 'medical model', and only very superficially acknowledged quality-of-life concerns and the subjective preferences of users. When these were taken into account, analogues appeared to be extremely valuable indeed. These newer types of insulin act on blood sugar in highly precise ways. For example, Novorapid – Novolog in the United States – is Novo Nordisk's flagship rapid-acting analogue. It begins to work after fifteen minutes or so, and peaks after only one hour – a far shorter and more intense action profile than standard soluble insulin.[50] By contrast, Sanofi's Lantus, an ultra-long-acting version, remains active for nearly a full twenty-four hours but has only a very modest peak.[51]

As a result, they allow for the intense fine-tuning of control, rapid corrective doses, and a highly adaptable regimen that can be changed on the fly as needed with far more precision than older styles of insulin allow. In the context of 'patient-led' therapy, particularly where it incorporates 'intensive' approaches to treatment, this can be a great boon, and it should come as no surprise that Kinga Howorka – discussed in Chapter 3 – strongly endorsed them.[52]

Unsurprisingly, while their impact on clinical outcomes like HbA1c has long been debated, the increased flexibility offered by analogues proved highly popular with people with diabetes.[53] This was per-haps most starkly demonstrated in Germany. In 2006, the country's Institute for Quality and Efficiency in Healthcare (IQWiG) recom-mended against their use entirely, arguing that the benefits appeared marginal at best. The decision prompted public outcry. Some of this came from doctors, who did not appreciate being told what they could and could not prescribe. Importantly, though, much of the furore came from people actually using insulin. Regardless of the lack of conclusive evidence that analogues offered any meaningful clinical benefit over older varieties, their subjective expertise said otherwise, and they had enthusiastically embraced the new versions.[54]

In the end, the episode was resolved without much difficulty and in a fashion that, for once, satisfied everyone bar the pharmaceutical industry. The authorities refused to budge without hard proof that analogues were actually superior, which, of course, they understood in 'medical model' terms. The manufacturers, however, had not conducted any blinded studies that might demonstrate this – having successfully convinced the relevant regulatory bodies that such research was impractical. Realizing that they were at an unsalvageable stalemate they relented, decided that a modest profit was preferable to no profit at all, and reduced the price of analogues in Germany to match that of ordinary biosynthetic insulin. Once the costs came down, the IQWiG saw no reason to maintain its opposition.[55] Where the political will exists, the price of medication can always be made flexible.

In the United States, however, no such state intervention has occurred. There is, as a result, very little incentive for insulin manu-facturers to ensure affordable prices. Unsurprisingly, this has led to spiralling costs. Humalog, Eli Lilly's flagship short-acting analogue, cost $21 per 10 ml vial in the United States when it was first introduced in 1996.[56] Were this adjusted for inflation alone, it would now cost about $40. In fact, the same amount goes for $274.70.[57] The vial, how-ever, is effectively identical to what existed a quarter of a century ago. It even comes in virtually the same packaging that it did in the 1990s![58]

Nonetheless, some argue, the whole notion of a 'crisis' is flawed – insulin is not expensive to buy in itself, only newer analogues are, and they are not the only option. Walmart, for example, sells it for only $24.88 per vial – less than a tenth of the price of Humalog!

Notwithstanding the fact that this might not exactly seem cheap for those already struggling to get by, it does seems a good deal better than opting for more expensive, brand-name analogues. In practice, however, this is not the simple solution that it first appears, and, particularly for the most desperate, this advice can be outright dangerous.

It is true that insulin can be bought for $24.88 from Walmart.[59] Is this a viable solution to the accessibility crisis, however? No. So-called 'Walmart insulin' is a colloquial term for preparations sold as part of the retail giant's 'ReliOn' range. Until 2021, the company offered only three basic types – Novolin R, Novolin N, and a premixed 70/30 version of both, manufactured on their behalf by Novo Nordisk. Novolin R is basic, soluble insulin – the 'R' stands for 'regular'. It is a biosynthetic version of the same thing that was developed in Toronto in the early 1920s, and is very similar to Eli Lilly's Humulin. Similarly, Novolin N is a biosynthetic version of NPH, an extended-action formula developed in Denmark all the way back in 1946.

So 'Walmart insulin' is old, but what is the problem? The oft-repeated claim that these older varieties are *inherently* harmful for the people using them is not strictly true.[60] Neither is the idea that they would necessitate a return to the poor prognoses, strictly scheduled timetables, and dietary prescription that had characterized much of the twentieth century.[61] As the Düsseldorf model demonstrated, it was perfectly possible to implement effective MDI treatment with the kinds of insulin available in the late 1970s and early 1980s. It would, however, be very difficult to achieve anything approaching the flexibility permitted by analogues with these varieties, so it would be absurd to suggest that they could ever function as a long-term solution for those in need. In theory they could keep people alive and (relatively) well in extremis, but used in this kind of desperate context they *do* pose a real danger.

The problem, here, is that older insulin acts on blood sugar in a very different way to analogues. If someone using a Humalog/Lantus basal-bolus MDI regimen, for example, moved to a soluble/NPH one, they may assume that they are exchanging like for like, and continue to inject and eat as they always have done. The basic principle of treatment is the same, after all. This, however, would almost certainly lead to highly unstable blood sugar because the two combinations act on the body in completely different ways. Soluble insulin works for far longer than Humalog, and NPH for much less time than Lantus, with a significantly more pronounced peak. If these differences are not accounted for, disaster looms, with potentially fatal results.[62]

This can, of course, be adjusted to. With sufficient knowledge and support there is no reason that someone could not learn to effectively use 'Walmart insulin', albeit at some cost to their overall quality of life. When certain organizations – many of them in the pocket of the pharmaceutical companies – discuss moving onto these older versions, however, they often fail to appreciate the reality of the situation for those they address.

The diaTribe Foundation, for example, is a San Francisco-based non-profit organization with the stated mission to 'improve the lives of patients through advocacy and social change'.[63] diaTribe is sponsored by an array of pharmaceutical concerns including each of the 'Big Three' insulin manufacturers, and with good reason: their individualist approach to the crisis steers well clear of suggesting wide-ranging legislative solutions that might threaten to work.

In one article published on their website, they address the issue of accessibility, and attempt to provide advice for those who may, for whatever reason, have found themselves uninsured or otherwise unable to afford their prescriptions. One potential solution, they suggest, is to use 'Walmart insulin'. diaTribe does emphasize that Novolin R and N are very different to the analogues people may have become accustomed to, and that caution should be employed when switching. Predictably, however, their position misreads the room entirely by suggesting that, before anyone does so, they should discuss the decision at length with their physician.[64]

This is completely divorced from the real experience of people who turn to 'Walmart insulin' out of desperation – people who may have no insurance at all. This is not a group that can always expect to be able to afford to visit a doctor for in-depth advice, nor pay for the SBGM equipment that might allow them to transition carefully and safely with strict self-observation. More often, they pick up their insulin and try to muddle along as best as they can. Some may find that this works for them, but moving directly from one type of insulin to another can have devastating consequences if it is not done with great care.

There is a historical precedent to this. The introduction of biosynthetic 'human' insulin in the 1980s was marred by deep controversy. In part, this originated in the (reasonable) suspicion that it actually offered little benefit, and primarily served to maximize corporate profits.[65] In his autobiography, for example, former British Labour party councillor Tony Huzzey – who developed diabetes in 1950 at the age of 12 – echoed many others when he argued that 'these developments were not introduced in any way whatsoever for the benefit of the diabetic but were solely for the benefit of the drug companies . . . whose profit per prescription soared'.[66] Economics aside, however, by the late 1980s some users were reporting that they no longer felt the familiar symptoms of hypoglycaemia until their blood sugar had already fallen to a dangerously low level, while others claimed that the character of the warning signs had changed. As a result, people were lapsing into unconsciousness before they had a chance to respond – a dangerous situation indeed.[67]

In many cases, individual people with diabetes had little say in switching to biosynthetic insulin. In the late 1980s manufacturers began to withdraw animal-derived types from the market, leaving them with little alternative. As one 1989 *BMJ* editorial highlighted, while only around 6% of what was prescribed in Britain in 1985 had been biosynthetic, 'at least three quarters' of the country's insulin-requiring population now used it.[68]

'Human' insulin is a very different thing to that made from pig or cattle pancreases – the kind most at the time were accustomed to. Unsurprisingly, it interacted with their bodies in new and unpredictable ways. While it soon became the standard type prescribed, and most people did learn to work with it, this clearly demonstrates the dangers of switching quickly from one formula to another, and often with minimal tailored guidance as to how to do so safely. There should be a lesson in that for those who continue to insist that 'Walmart insulin' represents an easy solution to the accessibility crisis.

In June 2021, Walmart announced that it would begin to offer Novolog as part of its ReliOn range – the first, and so far only, analogue that it sells. At first, this may seem a welcome development, and at $72.88 per vial it is certainly much cheaper than the branded version.[69] Nonetheless, this is hardly the revolutionary development the company seems to feel it is. No corresponding extended-action analogue is offered, leaving people reliant on Novolin N where they require one, as most using MDI do. More to the point, at three times the price of Novolin R, it is essentially positioned as a 'luxury' product available only to those willing to pay a premium – an additional cost the most vulnerable remain unable to bear.

Searching for work-around 'solutions' to the insulin crisis by looking to big business is, in any case, missing the point. Whatever small concessions these companies make, they are the problem. Novolog, the supposedly affordable analogue, continues to be sold at a significant profit margin, and for a far higher price than even branded varieties are in countries where the manufacturers know they could not get away with being so brazen.[70] The only sustainable way to address the insulin crisis is by permanently reducing the cost of insulin, ideally to zero, through lasting political action. This will always be unpalatable to private industry and its supporters in government, but there is no world in which equity of access can exist while both the manufacture and sale of essential pharmaceutical products are controlled by vested private interests which quite nakedly prioritize profit above all else.

It's Not Just The Cost

The insulin crisis is about more than simple availability. It is also about how the substance is prescribed and used. Even in countries with universal healthcare provision and secure supplies, the structures through which it is sometimes delivered can contribute to violent marginalization.

Prison is one obvious example of this. Those incarcerated – already more likely to come from significantly underprivileged and often non-White backgrounds – are rarely allowed to keep their insulin supplies in their own cells. They must either rely upon the institutional medical staff to perform their injections for them, or do them under close supervision.[71] While most prisons do have formal procedures in place for those on insulin treatment, these are not always followed consistently, and the potential for deliberate, punitive neglect is obvious.[72] In Oklahoma in 2017, Wayne Barnes, a jail administrator, was even prosecuted for his part in the 2013 death of Kory Wilson, a 27-year-old man with T1DM. Wilson had arrived at the institution on 16 June, but had never received a medical evaluation. When he asked to be given insulin, Barnes ignored him, and even after he started to become visibly ill accused him of faking his symptoms. He did not call for medical help for a full three days, and even then only after finding Wilson unresponsive on the floor of his cell. He later died of DKA. Barnes was sentenced to just over four years.[73]

The process of being arrested itself can also be highly dangerous. Like most people, police officers are often not particularly knowledgeable about diabetes, and sometimes do not appreciate quite how important insulin – or, in the case of hypoglycaemia, sugar – can be, sometimes over a very short period of time. This can put detainees at serious risk. If syringes and other equipment are confiscated, and the jailor remains deaf to complaints, they are left powerless as their blood sugar spirals out of control. In one 1993 study, for example, the authors pointed out that, between January 1989 and June 1991, forty-nine prisoners in New York City were collectively hospitalized with DKA fifty-four times. 70% of these cases occurred because they had not been given insulin after being arrested.[74]

Prison is a relatively niche example, but similar problems exist throughout society. Prescriptions must always be signed off by health-care professionals. They may even insist upon seeing their patients at an appointment before they agree to provide the necessary paper-work.[75] One (alleged) doctor on an internet question-and-answer thread, for example, described how she would sometimes withhold repeat prescriptions:

> I have had patients that refused to get labwork done or follow up and only wanted refills. At some point, I would refuse to continue this without seeing them because it would not be appropriate care ... Not getting follow up can be risky – for their health – and for me if anything happened as a result.[76]

Of course, the obvious counter is that – in the case of T1DM especially – not being able to access insulin at all might be a *lot more* risky! Whether or not they act upon it, this implicit authoritarianism is invoked more commonly than is comfortable. While finishing the edits on this book, for example, I received an invitation to a vascular clinic, which concluded by suggesting that 'it may not be safe to continue prescribing your medication [insulin] until you have been reviewed'. This may well have been a standardized message, but in the context of T1DM it could not help but feel a little threatening – it reminded me of how vulnerable I actually am, and how powerless I would be to do anything should the medical authorities choose to cut off my supplies. That people with diabetes must rely upon their doctors for repeat prescriptions that they would simply die without seems absurd in the extreme. How can this be read as anything but a strategy by which professionals work to maintain control over their patients?

Similarly, prescribing decisions are not always made on wholly medical grounds. In Britain, for example, Diabetes UK criticized the NHS in 2017 for restricting supplies of SBGM testing strips on the grounds of cost. Having enough of these is essential to managing diabetes, because regular tests are the only way of assessing overall blood sugar levels on a day-to-day basis – especially so with 'patient-led' treatment that involves frequent self-adjustment.[77]

In private healthcare systems like that of the United States, this barrier to accessibility is only amplified because provision is determined not only by doctors and healthcare organizations, but also by the commercial needs of insurance providers. According to one 1999 study, even as they were becoming ubiquitous in other parts of the world, only 2% of insulin users in the United States used pens in their treatment because, being more expensive than a traditional needle and syringe, they were often considered a 'luxury', and were as a result not covered by many basic insurance plans.[78]

Similarly, insurance policies often cover only certain types of insulin, which, given the substantial differences between them, can be a serious issue. This becomes an even more pronounced problem when they decide to change the varieties they will fund at short notice, sometimes forcing their customers to switch to a type they have no experience using – something that can, as we have seen, be both distressing and potentially dangerous.[79]

People with diabetes do not just need insulin to be affordable. They also need consistent access to it. This is a far larger project, and one that demands we not only engage with manufacturers, insurance companies, and government actors at the economic level, but also work to challenge the myriad power structures that might block or restrict access for whatever reason.

An International Problem

Much discussion about insulin accessibility centres on the United States. This should not come as any surprise, because it is the American experience that draws into focus more than any other the sheer scale and needlessness of the crisis. Despite its status as the wealthiest country on Earth, its citizens die in their thousands in the midst of plenty. Much like poverty, homelessness, and any of the other banal evils of capitalism, the lack of widely accessible healthcare in the United States has now become so normalized that many in the country look right past it without seeing anything untoward.

However, this is by no means a uniquely American concern. Canadians, for example, are often very proud of their supposedly 'universal' healthcare system, and there have been several occasions in which convoys from the south have made the news, crossing the border to buy insulin supplies for a fraction of what they would cost at home.[80] However, even in Frederick Banting's homeland, it can be cripplingly expensive.

While hospital treatment and doctors' appointments are free at the point of use in Canada, and emergency departments can provide those in desperate need with free medication, routine prescription costs are not generally covered by the health system for the working-age population. Canadians must, as a result, rely on private insurance schemes just like their American counterparts. Insulin does not cost so much in the north – it goes for around a tenth of US prices – but the burden of management can nevertheless mount up, especially when needles, syringes, insulin pens, and SBGM equipment, which are rarely covered by the public system, are taken into account, never mind expensive devices like CSII pumps or CGM technology.

The scale of the problem also goes well beyond the borders of North America. Rates of diabetes are increasing quickly throughout the world, with total figures accelerating at a staggering pace in the densely populated, rapidly industrializing economies of China and India. Pakistan currently has the highest global prevalence rate with a full 30.8% of the population living with one type or another, and many other countries

– particularly the Pacific island nations and the wealthy oil-states of the Arab world – have similarly astronomical figures. In Kuwait, 24.9% of all citizens are affected, and in Nauru 23.4%. The Marshall Islands, Mauritius, and Kiribati are not far behind at 23%, 22.6%, and 22.1% respectively.[81] The vast majority of these people have T2DM, as do the vast majority of people with diabetes anywhere. The worst impacted countries, not coincidentally, tend also to have rather high rates of obesity for a variety of complex demographic reasons.[82]

In the wake of the Second World War, for example, American influence in the Pacific rapidly led to shifts in the food culture of many island communities – traditional fish and vegetable-based meals were increasingly supplanted by imported, shelf-stable foods. Today, Spam is as much a part of Polynesian cuisine as Mahi Mahi or breadfruit.[83] To no one's surprise, however, the transition has had public health consequences.[84] This is not the place for an in-depth analysis of *why*, precisely, each individual country has the diabetes prevalence that it does, but the need for accessible insulin worldwide is clear.

That some people live without access to adequate healthcare is always the product of political choice. In some cases, this is readily apparent. The United States' continued embargo of Cuba, for example, has led to major shortages on the island, and no few unnecessary deaths.[85] More often, however, the causes are more banal and all the more troubling for it. In Lebanon, for example, prices have spiralled thanks to the government's policy of austerity, spurred on by corruption and apathy.[86]

Insulin is not particularly expensive or difficult to produce, but it is an effective money-spinner. As it turns out, people will pay quite a lot when the alternative is death. Buying an existential necessity, consumers have little power to refuse any price-tag, artificially inflated though it may be.

Private companies will always seek to maximize profits. In the United States, they are directly facilitated by the government in doing so, not least because corporate lobbying groups can wield real power over policy. While still highly expensive relative to the income of the local population, insulin in many poorer countries remains considerably cheaper than it is in America.

The United States serves as a clear demonstration of the consequences of allowing near-unfettered capitalist interests to govern the

delivery of healthcare according to neoliberal principles. That manufacturers charge so much for insulin there is not, as many understand
it, some strange American aberration, evidence of corporate 'greed'
ruining an otherwise functional system, but rather the opposite. Insulin
costs so much in the United States *because* it is the richest country on
Earth, because the manufacturers know that people will pay even if it
puts them into crippling debt, and because the federal government will
do nothing to meaningfully stop it.

Global accessibility, however, cannot be achieved through any single-
issue campaign centred on insulin alone. Even for those with diabetes,
affordable insulin does not mean affordable SBGM equipment, or needles, or treatment for the variety of potential complications that hang
like a shadow over everyone affected. But everyone needs healthcare
eventually, rich or poor. Access to insulin cannot, therefore, be separated from a broader demand for access to comprehensive universal
healthcare. Achieving such a thing, however, necessitates sharp critical
engagement with the socio-economic ideologies that shape the world
in which we live.

Conclusion: Insulin for All?

A century after it first began to save lives, insulin serves as a stark warning about the dangers of neoliberal capitalism. As of 2022, it is manufactured entirely by commercial interests. Even Connaught Laboratories, the University of Toronto-affiliated non-profit originally contracted to make it, was fully privatized in 1986 during a wave of economic reforms in Canada.[1] What remains is now owned by Sanofi, one of the 'Big Three' insulin manufacturers.

Neoliberalism has done more than hand responsibility for most of the planet's insulin supply to private industry. It has also exerted a subtle cultural influence, reinforcing the individualistic moralism that has always coloured the way people with diabetes have been perceived both by medical professionals and society at large, and in doing so has fuelled harmful public misinformation. In popular discourse, it is still common to see diabetes – whatever the type – erroneously attributed to overconsumption, particularly of sugar. It is not difficult to find examples of people describing, for example, a particularly rich dessert as 'diabetes on a plate', or jokingly claiming that some particularly sweet item of food 'gave them diabetes just by looking at it'.[2] Worse, this is not limited to a few ignorant comments, and depictions in the mainstream media frequently involve similar inaccuracies.

In some cases, this is the product of simple laziness, or perhaps apathy, in production. For example, in 2015 the third season of BBC drama series *The Syndicate* featured one character with T1DM who became disoriented due to a bout of hypoglycaemia and yet was treated, apparently successfully, with a shot of insulin – the exact opposite of what would have been an appropriate response had the situation occurred in real life.[3] This simple misrepresentation could easily have been avoided with minimal research, and it does not seem excessive to expect that the show's creators should, ideally, ensure that they have at least a basic awareness of the things that they use as plot points. If someone were to come across a person experiencing serious hypoglycaemia and decide to inject them with insulin, having seen something similar work on television, the results could be catastrophic, and perhaps even fatal.

Technical inaccuracy, however, is only the tip of the iceberg. It is arguably even more harmful when media productions perpetuate negative, morally charged stereotypes about diabetes and insulin. This continues, unfortunately, to be highly common across mediums.

One of the more egregious examples in recent memory might be 2013's *Hansel & Gretel: Witch Hunters*, an action-horror adaptation of the classic Brothers Grimm fairy tale.[4] In it, the idea that diabetes is a condition of overindulgence serves as a central pillar of the plot. As a child, Hansel is force-fed sweets by a witch, who, intending to eat him, wants to fatten him up. While he escapes, the experience leads directly to his developing diabetes, and, as an adult, he must still take regular injections to keep it stable.

Hansel & Gretel is not intended to be taken seriously, and it revels in its own silliness. Nonetheless, reinforcing the idea that eating too many sweets causes diabetes, however jocularly, has real consequences. The character of Hansel might actually have been a valuable one. Many younger viewers living with diabetes might have appreciated the depiction of a fellow-traveller as fearless hero in spite of it. However, the film cannot escape the implicit moralism so familiar to its subject-matter. Hansel is cast as blameless because diabetes was forced upon him by the actions of the film's antagonist, but this narrative only works because it embraces the overconsumption premise. In the real world few children must deal with cannibalistic witches. Nobody forces them to eat sweets. The implicit message is that if someone does end up needing insulin in real life, they must have done something to warrant it.

This framing is not the sole preserve of nuance-light popcorn fodder. It also crops up regularly in otherwise considered media. Despite its largely amoral cast of obnoxious characters, for example, sitcom series *It's Always Sunny in Philadelphia* is, for the most part, a thoughtful and subtly progressive TV show that often engages productively, in its own irreverent way, with politically controversial subject-matter. However, in one plot arc from the show's 2011 seventh season, Rob McElhenney's Mac becomes overweight. He is subsequently shown eating a literal sack-full of takeaway food while simultaneously injecting copious amounts of insulin.[5] This is used as a throwaway joke – the delusional character fancies himself a tough-nut, and believes that he is 'cultivating mass' – and, in typical sitcom style, the whole thing is soon forgotten, but the implications are troubling.

It's Always Sunny in Philadelphia's depiction of diabetes involves more than an explicit association between obesity and T2DM. It also makes a clear moral statement. Mac needed insulin, the show suggests, because he allowed himself to get fat, and developed diabetes as result. Needing insulin is cast as a punishment for wanton indulgence, which the audience is encouraged to laugh at. That the character nonchalantly injects dose after dose while continuing to consume is used to indicate his hedonistic attitude, legitimizing his status as an appropriate object of derision.

While some would argue that criticism here is unnecessarily pious, it is essential. Depictions like this are instrumental in normalizing the narrative of an individual, largely self-inflicted condition associated with overindulgence and laziness, and, as a result, one undeserving of collective concern.

Clinical Issues

Morally loaded depictions of diabetes in the media shape public perception in harmful ways, but the implications of these stereotypes for medical practice are arguably even more dangerous. If the social consensus holds that people affected are, as a rule, irresponsible, then clinicians are encouraged to adopt a paternalistic view of them – as childishly unruly and, implicitly, of requiring strict supervision.

Persistently dismissive attitudes towards people with diabetes amongst healthcare professionals are not difficult to find, and they often parallel the negative stereotypes that remain ubiquitous in popular discourse. This is particularly the case amongst non-specialists. In 2019, for example, a lecture slide surfaced on social media that highlighted some survey responses given by student nurses when asked their opinions about working in diabetes services.[6] Amongst those who were not attracted to the idea, the reasons given were telling. One respondent admitted that they had 'no patience for people who cause themselves to become ill, lose limbs, and disregard their medication/diet regimen', before indicating that they would 'become overwhelmingly frustrated working with this group of patients all day every day'. Another agreed with that assessment, arguing that 'from what I've seen thus far, many of those who have diabetes are noncompliant and don't take care of themselves'.[7] Once again, the classic stereotype remains strong. People with diabetes are often 'non-compliant' and, as a result, exhausting to work with.

This kind of moralistic attitude is not limited to those early in their careers, who perhaps might develop more nuanced opinions with experience. One 2019 event in London organized by Urgo Medical, a private company specializing in wound care, for example, was advertised with the tagline 'To scare or not to scare?'[8] The intended programme included a discussion about the potential value of healthcare professionals employing 'motivational interviewing' or, less euphemistically, 'scare tactics' in consultations involving people with diabetes-related foot ulcers. When an advertisement for the event was shared on Twitter, the response was predictably hostile, but as one commenter pointed out, that it was ever published at all was troubling – serving as 'absolute proof that massive ignorance in some of the medical profession is alive & kicking when it comes to treating PWD [people with diabetes]'.[9]

This is, of course, not true of every practitioner, and there are, particularly amongst those who do end up specializing in diabetes, an increasing number who enthusiastically engage with their patients as complex, subjective actors, making considerable effort to treat them with dignity and respect. Reports like 2018's *Language Matters*, commissioned by NHS England, for example, highlight the importance of avoiding moralism and acknowledge that those with diabetes, whatever their HbA1c or other 'clinical' values, are usually doing the best that they can while living with an extraordinarily complex condition, and, more to the point, may have different priorities in management to the ones traditionally held by professionals.[10]

While publications like *Language Matters* are undoubtedly cause for optimism, there is a reason that they needed to be written in the first place. Even amongst those who work directly with people with diabetes, unhelpful attitudes remain disconcertingly common, and this has serious implications for the relationship between them and their patients.

In 2019, for example, Laura Marston, a well-known American diabetes activist, was highly frustrated following an appointment with her doctor. 'Went to my PCP [primary care physician] yesterday and had full bloodwork done', she reported. 'To the doctor who noted in my labs that my glucose tested at 106 [mg/dl – approximately 5.9 mmol/l] was 'elevated', go fuck yourself.'[11] A blood sugar value of this level would not be out of the ordinary even amongst the general population, and Marston's incredulity was understandable, but she was also subtly making a deeper point. By dismissing the words of the offending professional in such uncompromising terms, she is taking an ideological stand – highlighting the reductive paternalism she felt was at work while reasserting herself as a distinct moral actor in the framework of management.

Her example is hardly unique. Renza Scibilia, a member of the Australian advocacy community, is similarly forthright. The same year, she uploaded a photograph of a table used in her diabetes clinic to assess HbA1c values. Against each figure, the document listed corresponding 'grades', descending from 'Outstanding A+', to 'Very Poor E'.[12] Predictably, Scibilia soon discovered that she was far from the only person to notice similar. One person even shared a photograph of another almost identical document from the UK.[13] Again, it listed a variety of HbA1c figures alongside value-laden terminology such as 'good' and 'poor', and it was supplemented with pieces of advice encouraging the reader to 'improve food choices' and 'take much more care with tests'. A satisfactory result is annotated by hand with the phrase 'well done'. As Scibilia pointed out, 'it's like a fucking report card'.[14]

Perhaps most telling of all, however, is an appended passage which states 'Please note if your HbA1c is >10% you will be seen in clinic in 6–8 weeks, if it is >14% you will be admitted for re-education of your diabetes management'. The short paragraph is not particularly subtle in its meaning. It seems to imply quite plainly that those attending the clinic are *expected* to maintain a level of control that the staff considers adequate, and, if they do not, that they are evidently approaching their treatment in the wrong manner and must be disciplined as a result.

Suffice to say, the response to this document was not kind. Commentators criticized its moralistic tone and its failure to incorporate any acknowledgement of subjectivity, while, most importantly, questioning what right exactly, if any, professionals had to make value-judgements about the lives of those under their care.

These examples portray a profession still regularly beholden to de facto paternalism even while it pays lip service to 'patient-centred' care. Doctors know, supposedly, how best to maximize health according to the narrow view of the 'medical model'. In the context of diabetes, this is often debatable, but that is beside the point.[15] More important is that they frequently continue to invest this framework with coercive moral value, and in doing so dismiss their patients as complex subjective actors.

'Good Care'

While much of the continued paternalism visible in diabetology can no doubt be attributed to a long tradition of clinical authoritarianism, some of it is likely born of good intentions. As early as 1980, some professional commentators had begun to decry the 'physician who merely spreads an array of vendibles in front of the patient' and asks them to choose, apathetic to the outcome.[16] By the twenty-first century, however, the hegemony of neoliberalism seemed, in practice, to promote just that. In this context, far from emerging out of a desire to control individuals for its own sake, at least some paternalism can be read as a protective response to an ideological framework that threatened to leave them more vulnerable than ever.

One good example of this can be found in philosopher Annemarie

Mol's 2008 book *The Logic of Care*. In it, Mol addresses the prevalence of neoliberal attitudes in healthcare delivery and reflects upon the implications of recasting the patient as a rational consumer according to market principles. Using T1DM as a case study, she argues that this framework – which she describes as the 'logic of choice' – has, in effect, often eroded 'existing practices that were established to ensure "good care".'[17] Mol does not, however, stop there. She also outlines an alternative – the eponymous 'logic of care' – that 'starts out from the fleshiness and frailty of life'.[18]

The criticisms of neoliberal healthcare levelled in *The Logic of Care* are valid and convincingly argued, but Mol's manifesto for a new framework falls into a familiar trap. Within her 'logic of care', the idealized professional has an obligation to ensure as best as possible that the individual for whom they are responsible 'looks after themselves', which does not necessarily entail 'going along with them'.[19] Their job, in addition to the practical business of medicine, should, she believes, be to approach their patients at a moral level, and nudge them towards making 'good' choices. This, however, poses a problem. Who decides what a 'good' choice actually is? This is a fundamentally subjective question. For a physician, this will usually be perceived primarily in 'medical model' terms, but this may not align with the subjective needs and values of the person being treated. This seems a somewhat awkward position, and Mol does come close to articulating the problem, acknowledging that 'it is not always clear what to count as "improvement"' in treatment. She quickly brushes past this, however, to claim that 'despite these complexities, in one way or another, unstable blood sugar levels are bad'.[20]

While it is likely that many (and perhaps most) people with diabetes would agree with this, her vision implicitly privileges a particular kind of thinking and, in doing so, works to reinforce medical authority. While Mol does not believe that anyone should have the power to *compel* their patients to act in any particular way, she does believe that, where they are not acting in their own best interest – as defined by the physician – it is prudent to step in to persuade them to change course.

In practice, then, the 'logic of care' cannot help but collapse back into a de facto soft paternalism. It is difficult not to be sympathetic to Mol, and to physicians who try to oppose the neoliberal project in this way. This is a very difficult circle to square. But then we are back to the beginning: how *should* diabetologists interact with their patients? Is there another option? A framework of care that is neither apathetic nor subtly coercive? Perhaps there is.

Mol's 'logic of care' has many similarities to the 'deliberative' model of the physician–patient relationship outlined by Ezekiel J. Emanuel and Linda L. Emanuel in 1992:

> In the deliberative model, the physician acts as a teacher or friend, engaging the patient in dialogue on what course of action would be best. Not only does the physician indicate what the patient could do, but, knowing the patient and wishing what is best, the physician indicates what the patient should do, what decision regarding medical therapy would be admirable.[21]

Like the 'logic of care', the 'deliberative' model is well intentioned. Similarly, however, it places the physician in a position of moral authority over their patient. 'Good' outcomes are defined by the professional – if their patients disagree, they are, implicitly, making a mistake, and should be strongly encouraged to reconsider their position.

Emanuel and Emanuel, however, describe an alternative. The 'interpretive' model they outline also rejects the cold, uncaring sterility of absolute, unguided 'choice'. It invests the doctor with a role beyond that of a technician. However, it also asks that they avoid any attempt at moral coercion:

> The interpretive physician assists the patient in elucidating and articulating his or her values and in determining what medical interventions best realize the specified values, thus helping to interpret the patient's values for the patient.
>
> According to the interpretive model, the patient's values are not necessarily fixed and known to the patient. They are often inchoate, and the patient may only partially understand them; they may conflict when applied to specific situations. Consequently, the physician working with the patient must elucidate and make coherent these values. . . At the extreme, the physician must conceive the patient's life as a narrative whole, and from this specify the patient's values and their priority.[22]

The 'interpretive' model fully incorporates the subjective values of the person undergoing treatment into every medical decision. It does not, however, leave them to the wolves. Neither displaying apathy nor coercive authority, it understands the doctor's role as akin to a 'counsellor, analogous to a cabinet minister's advisory role to a head of state'. Their job is not to convince their patients of anything, but to work out how to assist them as well as possible by their own definition.

As Emanuel and Emanuel make clear, however, this does not mean simply taking everything they say at face value. As in the 'deliberative' model, critical engagement is important. If someone with diabetes were to say, for example, that they simply did not care about their blood sugar levels and were happy to accept the many risks entailed by that position, the 'interpretive' physician would not try to actively persuade them to rethink their stance. They might certainly, however, ask pressing questions in an attempt to find out why they feel as they do. Through this process, the professional works to promote intense self-reflection. They encourage their patient to think deeply about themselves, their beliefs, and the implications of their condition, acknowledging and working to resolve any contradictions that emerge. In this case, perhaps, the person might come to realize that they *do* actually care about their long-term health, but are worried that protecting it might imply lifestyle sacrifices that they are unwilling to accept. Once this is established and a more nuanced picture emerges, it is possible to begin to collaboratively construct a management plan that takes into account all of these complexities. What are the actual priorities of the person being treated? Are they mutually compatible? If not, which is more important? What measures of risk and sacrifice are acceptable?

The key point here is that the professional does not seek to influence, but rather to help the patient understand and articulate their own values, before then working to implement treatment in a way that best accords with them. Under an 'interpretive' framework, they are not robotic, and they need not leave their personal values wholly at the door. Challenging discussion might even help facilitate their patients as they attempt to pin down and articulate their subjective position. What is important is that the doctor should never consider themselves 'above' those they treat – as worthy of judging them and, perhaps, of finding them lacking.

The 'interpretive' model therefore seems, at present, to be the most desirable approach on offer to diabetology. It is not, of course, perfect. It would, for example, be easy to allow intensive value-interrogation to lapse into de facto moral coercion, and professionals would always have to be careful not to overstep the mark. Nonetheless, implemented effectively, this framework always at least tries to respect the values of the individual and tailor management accordingly. It is certainly how I would choose to be treated, were I given the choice.

Having satisfactory relationships with clinical staff is important. Where they are strained this can be highly stressful, and can add to the burden of management considerably. Nonetheless, this is of limited day-to-day relevance. People with diabetes spend a tiny fraction of their lives interacting directly with professionals. However much their doctors might try to exert paternalistic power, they are always free to disregard instruction should they so choose. In this sense they have never been passive recipients of care, and today, particularly in the age of social media, they have never been more collectively assertive.

Technology from Below

In 1978, Stanley Joel Reiser expressed concern that, for all of the clinical benefits they promised, the machines that he was witnessing become ubiquitous fixtures of medical practice might come with dark implications. As more 'objective' tests became accessible, he worried that the human element of medicine would disappear – the real patient under examination ignored in favour of the collective data produced by the technologies that surrounded them. Such machines seemed almost a metaphor for the worst of dismissive, authoritarian practice, reducing their subjects to their naked biology.[23]

Reiser was not alone. In 1988 James Trostle made the observation that technology in medicine often served to reinforce physician control over healthcare and rendered their patients ever more passive by reducing the amount of labour required of them in treatment:

> Recently we can see medical technology used to reduce the need for patient initiative. For example, long-acting medications are recommended over short, injections recommended over pills, and

office-centred interventions valued over those in the home. These interventions may be effective, but they risk turning patients from responsible subjects into responsive objects.[24]

While innovations like extended-action (and later pre-mixed) insulin were highly inflexible and did dissuade individual engagement with therapy, the more recent history of diabetes and insulin might give both Reiser and Trostle some cause for relief. 'Patient-led' management of the kind frequently used today acts to reverse this trend by – following Trostle's argument – transforming individuals back from 'responsive objects' to 'responsible subjects'.

Similarly, while the visible technologies involved in treatment – needles and syringes, CSII pumps and, increasingly, CGM devices, for example – can sometimes mark a person as 'different' and contribute to social isolation, they can just as often act as a vehicle for sometimes quite radical subjectivity, and through it provide avenues for the bold reassertion of individual humanity.

Sometimes, the meaning of 'otherness' can be subverted. Margaret Howie, for example, did not often mention her diabetes to people until she knew them reasonably well after she was diagnosed in 1975. In personally meaningful moments, however, she was able to use the context of insulin therapy to her advantage. At the beginning of her relationship with her eventual husband, her needle played an important role:

> I kind of thought enough's enough, you know? I'll just tell him and if he goes, he goes, and if he doesn't, he doesn't. Well, it turned out he'd been out with a diabetic when he was at school anyway so he had half a notion of what was what . . . I think he took me for a picnic or something and . . . we were sitting on the beach. I just took my thing out and did my injection. He's like 'Oh, right!'[25]

While it is clear that she felt considerable anxiety about the barrier her condition might create between her and her partner, Howie's decision to uncharacteristically perform her injection in front of him was a genuinely intimate act. By allowing access to part of her life that she otherwise went to considerable lengths to hide, she risked rejection, but she also turned the danger on its head. It was he, not her, who became the subject of judgement. By gauging his reaction, she could determine their ultimate compatibility.

This subjectivity is not always expressed in such private ways. Today, for example, it is not unusual for people with diabetes to explicitly reject the notion that their condition is anything to be self-conscious – let alone ashamed – about. It is simply a part of their life – an integral part of their humanity, not something that threatens it. Taken further, this attitude completely reorients the way visibly worn technologies of diabetes management might be perceived. These physical markers can no longer be read as unwelcome evidence of difference divorced from the 'real' person upon whom they are located, but are instead reclaimed as part of that body in their own right. They do nothing to undermine the humanity of those using them, but are wholly bound up with it. As extensions of the body itself they can even be cast as something to be celebrated – integral parts of the individual's subjective totality, without which they would be someone else entirely.

This attitude can be seen frequently today, particularly on social media. It is now very common to see photographs in which the subject makes no effort to hide insulin pumps, CGM technology, or the tubes and wires that go with them. These objects are presented either as an irrelevance or, in some cases, even cast as objects of beauty – as inextricably linked to, and part of, the beauty of the individual themselves. There are even a number of sex workers with diabetes who sell suggestive, and sometimes pornographic, images of themselves, all while defiantly wearing the devices that keep them alive.[26]

That many of those sex workers are selling content for the explicit purpose of affording essential medical supplies like insulin is troubling, but nonetheless these examples demonstrate clearly how medical devices, quite apart from being cold and dehumanizing, can actually, in the context of insulin therapy, be repurposed as vessels through which the subjective humanity of the user is amplified, not fractured. In these online spaces, this can be particularly powerful as people with diabetes act to reclaim control of the narrative about their condition, challenging common stereotypes as they do.[27]

The free exchange of information permitted by social media can also facilitate radical engagement with the technical business of diabetes management. Functioning as fully horizontal spaces in which traditional authorities have little coercive power, platforms like Twitter allow for the development of extensive peer support networks.

People with diabetes engaging with their treatment above and beyond anything conceived of in the clinic is not a new development, but social media has permitted both the scale and visibility of lay innovation to reach unprecedented heights. While for most this is a fairly informal process involving the exchange of subjective expertise, a minority have taken it one step further to directly take on the mainstream healthcare establishment at its own game.

Frustrated at the perceived slow progress of research and development in diabetes technology, some have decided to do it for themselves. After all, 'patients have – medical domain expertise, device security expertise, tech expertise. They are innovators, engineers, enablers, documenters, communicators, educators'.[28] Why, then, shouldn't they?

The most significant example of this lay engagement is a loose collection of individuals who have come together under the Twitter hashtag #WeAreNotWaiting. As a group, they work towards developing new approaches to insulin therapy, debate, and often act to defend one another by communicating with the traditional scientific authorities. In short, they represent a distinct research community comprised of people with diabetes – many of them without any formal medical training whatsoever.

One of the most striking successes of #WeAreNotWaiting has been the open-source development of code designed to force CSII pumps to respond dynamically to data provided by CGM technology – something that they had never previously been intended to do – via the exploitation of software vulnerabilities in older devices.[29] 'Looping', as the technique is colloquially known, uses algorithms to, in theory, dramatically reduce the amount of labour necessary in day-to-day management. In many respects, the technology functions in the same way *Balance* hoped 'artificial pancreases' one day might in the late 1970s.[30]

The members of the #WeAreNotWaiting community are able to do what they do because they primarily tinker with existing technology. The collective does not create any new devices, and is as a result able to bypass many of the approval and licensing requirements faced by more traditional research and development organizations.[31]

Unsurprisingly, even while #WeAreNotWaiting activists now host conferences and publish their own research, the movement has been met with considerable hostility from establishment interests.[32] While several formally licensed 'looping' technologies have since been released, these are expensive, and they tend to be less flexible than the DIY variants that many users have chosen to stick with.[33] For their part, some healthcare professionals have begun to engage with this kind of lay innovation, but they continue to do so in an uncertain and deeply cautious manner.[34]

#WeAreNotWaiting serves as clear evidence of the limitations of the 'expert patient' in its neoliberal interpretation. Anthropologist Samantha Gottlieb, for example, has succinctly expressed the way lay innovators in T1DM take the concept much further than many in traditional positions of authority are comfortable with:

> Their 'super engagement', which in theory, the FDA might fanta-size about in other patient populations, however, has not always been received enthusiastically by clinicians, regulatory, or commercial entities. The FDA's promises to include patients draw on existing clinical models of research and development, but they do not trouble the fundamental categories of what is possible. Perhaps it is unrealistic to expect the U.S. drug regulatory agency to restructure entirely, and, thus, the slow progress to include patients is still notable. But it is not just the anthropologist's fantasy that there are alternative worlds possible. The interventions the open source T1D communities have accomplished reveal concrete methods for patient expertise and knowl-edge to lead health technology. Few disease conditions present as many decision-making moments for patients, and thus, T1D is unique; yet, this community also introduces a paradigm shift. Their enactment of 'engaged' patients illuminates how regulatory fantasies of empowered patients neglect patients' own versions of participation. The FDA imagined fantastical empowered patient is a compliant patient, rather than the patient-as-actor-creator and disrupter.[35]

In short, online diabetes communities – #WeAreNotWaiting chief amongst them – expose the palpable contradiction at the heart of the neoliberal 'expert patient'. Expertise and empowerment are often respected only where they do not threaten the established power structures of medical orthodoxy. Furthermore, they often work to challenge the ideological assumptions of neoliberalism in a deeply meaningful way. While the original developers of DIY 'looping' technology might be read as engaging in quite radical individualism to ensure that their treatment met their subjective needs, this implies no rank self-interest.

The innovations that have resulted have, almost without fail, been enthusiastically mobilized to the collective good. One vital function of the #WeAreNotWaiting movement is that it acts as an information exchange network. The fruits of its labour are made available to everyone who might benefit free of charge.[36] Perhaps it is little wonder that the for-profit companies attempting to commercialize similar technology feel more than a little threatened.

Technology from Above

In 2021, I was invited to contribute to *Babbage*, *The Economist*'s science and technology podcast. The producers wanted to record an episode about insulin to commemorate the hundredth anniversary of the first experiments in Toronto, and they asked me, as a historian, to give an overview of the discovery while also commenting on accessibility and the implications of more recent research being conducted by the pharmaceutical industry. From the outset, it was clear that they were interested in exploring the idea that technology might have a role not only in making diabetes less of a burden in the future, but also in solving the insulin crisis itself.

While innovations like CSII pumps and insulin pens have had a considerable influence on diabetes management, they are the exception, not the rule. Most novel technologies have failed to meet expectations. Non-invasive methods of insulin delivery are a prime example of this. Pills, nicotine-patch style infusion technology, and inhalers have all been proposed as possible alternative methods of administration at various points since the turn of the millennium, but their impact has been limited.[37] One of the reasons why these technologies have failed to get off the ground – beyond technical complexity – is simply that, in practice, many of them do not actually hold much appeal to people with diabetes. Pfizer's Exubera inhaler is a clear example of this. When it was released in 2006, it failed to meet even a fraction of the sales targets expected of it, and, in what was later described as one of the greatest commercial flops in modern biotechnology, was withdrawn after only a year on the market, costing the company an estimated $2.8 billion.[38]

Pfizer tried to deflect blame for the disaster – in Britain, for example, they accused NICE of failing NHS patients by advising against its routine prescription – but the reaction was muted across the globe.[39] One commentator later derisively quipped that is was 'hard to fathom how a company that sells $60 billion worth of pharmaceuticals per year could misjudge a target market as badly as Pfizer did'.[40] Had the multinational paid attention to the people they hoped would use their shiny new device, they might have saved themselves the money. When NICE assessed the technology, they did, in fact, consult people with diabetes, and found that, as a whole, they had no reason to recommend the inhaler over injections or CSII treatment. The device was criticized for its bulky size, its inability to reliably deliver precise dosages, and for its numerous respiratory side-effects.[41] The report also highlighted that its incompatibility with long-acting insulin types – an important part of basal-bolus treatment – meant that regular shots would usually be required anyway.

Exubera was built on a false assumption – that the pain of injections themselves were a serious quality-of-life issue, and that many or most people who required them would enthusiastically embrace any technology that presented an alternative. As the lay representatives consulted by NICE reiterated, however, this was simply not true. 'Using injected insulin', the report argued, 'is not usually a major concern for the majority of people with diabetes.'[42] Injections are, for all but a tiny minority, not particularly onerous once they have become accustomed to.

While some physicians did express concern that a potentially useful technology was being resisted on essentially economic grounds, Exubera, despite Pfizer's bluster, failed to perform because it was an expensive and flawed solution to a problem that did not exist at anything like the scale that they thought it did.[43] It was not created to meet the needs of people using insulin, but rather to meet what Pfizer assumed those needs were.[44]

The complexities – and subjectivities – of insulin treatment ensure that the only people equipped to bring real insight to the table are those who undergo it every day. That is why groups like #WeAreNotWaiting have achieved some great breakthroughs where major pharmaceutical companies have not, with a fraction of the resources.

Of course, if insulin were no longer required at all, diabetes would be effectively cured. There is some reason to believe that, in the long term, this might not be the pipe dream that it first appears to be. Islet cell transplants, for example, have shown some qualified success in restoring function since the early 1990s, but, relying on scarce donor organs and requiring lifelong immunosuppressant medication, they are limited in scope.[45]

Other technologies look more promising. Cell encapsulation, for example, involves implanting stem cells into the body, shielded by a synthetic pouch that allows essential nutrients and hormones to pass in and out. They then mature into functional islet cells, which begin to make insulin hidden from the body's immune system. These devices – for example Viacyte Inc.'s VC-01 – are still in the trial phase, but could in the future theoretically provide a functional cure without any need for immunosuppressant medication.[46]

Cell encapsulation technology looks impressive on paper, but we should nonetheless remain cautious about any promised 'magic bullet' that will, once and for all, render diabetes an irrelevance. Ask almost anyone who has lived with the condition for any significant amount of time and they will tell you the same thing – that this particular Holy Grail has been supposedly just around the corner for decades. In reality, the basic premise of insulin therapy today has not changed since 1922, and it will likely remain the same for years to come.

Even if some new technology did work as a functional cure, by itself this would do nothing to address the global insulin crisis. A cure is only a cure to those who can afford it, and none of the major pharmaceutical companies look set to change policies that continue to make them vast sums of money, unscrupulous though they may be. Eli Lilly, for example, came to a $473 million deal with Sigilon Therapeutics, a pioneer in cell encapsulation technology, in 2018. True to form, the contract guarantees them an exclusive global licence for the finished product.[47]

Cell encapsulation technology is likely to be highly expensive if and when it becomes widely available. On the plus side, however, it may also be at least semi-permanent. Once the device is implanted, there is no reason to believe that it will not last for a considerable amount of time – maybe even for life. This, perhaps, explains why investment by the major pharmaceutical companies, while considerable, has been dwarfed by the money thrown at another new technology: glucose responsive insulin (GRI), colloquially known as 'smart' insulin.

GRI would not represent a functional cure in the same way cell encapsulation might, but it could transform the way that diabetes is managed while making it much less challenging to maintain stable blood sugar levels. Unlike standard insulin, which must be carefully measured, GRI, in theory, requires considerably less precision. It is designed to dynamically react to the current level of sugar in the blood and, in short, work only when its concentration is above a certain threshold. This almost completely averts the risk of hypoglycaemia, and means that so long as enough is injected, sugar levels should remain consistently within a safe range.[48]

Like cell encapsulation technology, GRI is still very much in the development stage, but while the concept was almost wholly the preserve of small, creative biotech start-ups and university-based research projects in the early 2000s, the pharmaceutical industry has started to take real notice. SmartCells Inc., the first company to seriously begin research in this area, was bought out by multinational Merck & Co. for over $500 million in 2010.[49] Eli Lilly, always quick to smell an opportunity, followed suit. In 2016 it acquired Glycostasis Inc., and, in 2021, agreed to a ~$1 billion takeover of Protomer Technologies – both organizations at the forefront of GRI development.[50]

There is one obvious reason why 'smart' insulin has prompted such excitement amongst manufacturers – it promises significant returns. Despite its clear advantages, GRI is, at heart, just insulin, and like any other variety, it needs to be injected every day. As the old adage goes, there is a great deal more money to be made in treating a long-term condition than there is in curing it outright.

While my comments on the topic were – perhaps unsurprisingly, given *The Economist*'s centrist political stance – cut from the final edit of *Babbage*'s insulin episode, looking to technology to provide any long-term solution to skyrocketing diabetes rates across the globe is doomed from the start.[51] If any great innovation is going to mitigate the harm, it is an international commitment to ensure access to adequate healthcare for all. Technology is nice, but in our present world it cannot help but feel like a distraction from the real issue – vast economic inequality.

Completing the Circle

Not long after Leonard Thompson received his first doses of insulin in early 1922, Frederick Banting had a vicious altercation with James Collip. Angry at the way he had been treated, Collip had threatened to walk, and to take the secret to purifying the group's pancreatic extract with him. He knew that the discovery – which, with some justification, he felt was essentially his anyway – could make him a very rich man, and, as a chemist not a doctor, making some money from it would not at the time have been considered particularly poor form.

Banting, however, for all his flaws, took his responsibility towards medicine as a collective human good very seriously indeed. He had always been uncomfortable with the notion of patenting insulin at all, never mind profiting from it. The first-hand accounts of what actually happened between him and Collip tiptoe around the specifics, suggesting that the fight was a lot more serious than anyone present felt comfortable committing to paper. Charles Best witnessed it, and in a 1954 letter recalled being 'disturbed for fear Banting would do something which we would both tremendously regret', so much so that he was forced to restrain him 'with all the force at [his] command'. Afterwards, he considered the chemist 'fortunate not to be seriously hurt'.[52] It is probably safe to assume that this was no polite disagreement.

This episode encapsulates perhaps the most immediately relevant lesson to be learned from the history of insulin. Banting was not a humble man. His endless self-belief frequently bubbled over into a palpable arrogance that many around him found obnoxious. Nonetheless, he clearly believed that insulin, and all medical innovation, should benefit humankind collectively. He was no profit seeker and he resented those who were. In 1922, he consistently refused to be courted by wealthy private interests, and this ethos persisted throughout his life. 'Insulin does not belong to me', he is often reported to have said in 1923, 'it belongs to the world'.[53]

Yet this simple substance is a privilege denied to many. There is no material reason why this should be so. We have the resources to produce as much insulin as everyone on Earth requires many times over. In fact, we probably produce close to that. Like the tons of food allowed to rot in locked supermarket bins a stone's throw from homeless camps, vast quantities of insulin end up sitting in warehouses and pharmacies, or in the homes of those fortunate enough to build a surplus, until it reaches its expiry date and is disposed of, all while many thousands across the globe are left disabled or dead for lack of it.

This occurs because almost all insulin production is controlled by a handful of powerful manufacturers who profit extravagantly from the status quo – Eli Lilly now routinely earns over $6 billion dollars annually. This state-sanctioned disaster capitalism, however, comes at significant human cost.[54] That each of the 'Big Three' companies are, at the time of writing, gleefully hosting celebrations to mark the centenary of insulin's discovery seems a cruel joke indeed.

When people criticize so-called 'Big Pharma', they often focus on the idea of 'greed' amongst the offending companies. Bernie Sanders, the Vermont senator and one-time presidential hopeful, for example, frequently uses this kind of language. Railing against the 'unregulated greed of the pharmaceutical industry at work', he condemns their 'profits before people, no matter the cost' outlook.[55] Sanders is famous as one of the more left-wing mainstream politicians in the United States, and his position is both widely shared and well-intentioned. However, it also reflects the hold neoliberal ideology has upon not only our society and governing institutions, but also the way we interpret the world.

The 'greed' argument presupposes that companies like Eli Lilly, Novo Nordisk, and Sanofi are somehow aberrations – that they, as individual entities, are 'bad apples', ruining the party for the rest of us and for more 'responsible' enterprises that seek only 'moderate' profits. The implicit message here is that more 'enlightened' corporate leadership, technocratic legislation, and civil discourse is the key to solving the crisis, but that the economic framework in which it has occurred is, if not desirable, at least immutable.

This, however, represents a dangerous failure of vision. Insulin manufacturers are not exceptions to the normal functioning of 'healthy' capitalism. They are capitalism in its most naked form. They ruthlessly exploit their customers to maximize profit margins not out of some vague, moralistic sense of 'greed', but because that is exactly what they are designed to do. Where they differ from most commercial organizations is not in their approach to business, but in the material conditions in which that business is conducted. They enjoy a captive, desperate, core customer base, virtually zero real market competition, and – in North America particularly – minimal legislative oversight. They are not 'bad' companies. Within the ideology of free-market neoliberal capitalism, they are extremely good at what they do.

If this unsustainable and cruel state of affairs is ever to change, a collective, internationalist response is necessary. There are some signs that this may not be the improbable dream it appears. People with diabetes, for example, often do what they can to assist one another. There is now what some have described as a significant 'black market' in insulin, though the term is a little misleading – money rarely changes hands.[56] Instead, when someone is identified as requiring emergency supplies, activists work to source what is required, often making extensive use of social media platforms to do so. There are now even dedicated websites – Mutual Aid Diabetes, for example – which can be used to directly request help when needed, most of them run entirely by volunteers.[57]

This informal system of community support cannot, of course, meet the needs of everybody, and it is certainly no replacement for fully legislated universal healthcare. Nonetheless, for some of those staring down the barrel of a gun, it has undoubtedly provided much needed breathing space, while the ethos it encapsulates is one of radical collective solidarity. Social media posts highlighting individuals in urgent need often receive significant engagement, and there are frequently numerous respondents – sometimes from across the globe – indicating their willingness to ship spare medication.[58]

This ethos also comes across in other ways. There are examples, for instance, of a productive overlap between accessibility advocates and the kind of grassroots technological innovators encapsulated by #WeAreNotWaiting. The Open Insulin Foundation (OIF), for example, comprises a group of self-described 'biohackers' with the stated goal of providing a viable alternative to reliance on the pharmaceutical industry. To this end, they are currently working towards the development of open-source insulin production methods which may one day allow small-scale batches to be made at the community level.

Originating in a community biology lab in Oakland, California, with only $16,000 of crowdsourced funding, the network has grown to be a truly international project, with collaborators across the United States, Brazil, Senegal, Cameroon, and Puerto Rico, and it is explicit in its mission statement. Directly alluding to the vast inflation of prices by the mainstream manufacturers, the group envisage 'a world in which communities in need have local sources of safe, affordable, high-quality insulin, and . . . can own and govern the organizations that produce the medicine they depend on to survive'. In short, the OIF believes that healthcare should be a human right, that all medicine should be open-source and community-governed, and that science as a whole should be democratized, not dominated by a handful of powerful organizations. 'We are taking over', says its founder Anthony Di Franco, 'the steward-ship of the medicine our lives depend on.'[59]

There is, surprisingly, some precedent for grassroots insulin produc-tion. When Eva Saxl fled Czechoslovakia with her husband Victor to escape the Nazis in 1940, for example, the couple ended up settling in Shanghai, one of the few places that continued to accept Jewish refugees with little restriction. Soon after arriving in China, however, she began to experience the symptoms of T1DM, and started insulin therapy. For a time, this was manageable. While it was under often brutal Japanese occupation, Shanghai was still a great metropolis and – because of its colonial history – was by far one of the region's most cosmopolitan and globally interconnected cities. For those who could afford it, insulin – most of it made by Eli Lilly – was readily stocked in local pharma-cies. After the Japanese attack on Pearl Harbour in December 1941, however, things changed rapidly, and it became exceptionally difficult to import supplies.

By 1943, Saxl was becoming increasingly anxious. In a desperate last-ditch gambit, she and her husband decided to try their hand at making their own insulin based on descriptions of the production process they found in medical textbooks. Making use of pancreases sourced at a nearby market, they began work in a small neighbourhood laboratory generously loaned to them by a local Chinese food chemist. Incredibly, they were successful. The murky-brown substance they came out with was probably not unlike that created in Toronto some two decades prior and, incredibly, it seemed to just about work.

The Saxls went on to establish a long-running production operation in Shanghai, providing vials of their home-brewed formula free of charge to those with diabetes throughout the Jewish ghetto, many of whom would almost certainly not otherwise have survived. Eva herself went on to become something of a celebrity in the world of diabetes and lived into her eighties, passing away in 2002.[60]

Those involved with the OIF would likely find much to admire in Eva and Victor Saxl. That they were able to manufacture usable insulin in such desperate conditions was a remarkable achievement in itself. The fact that they were able to produce enough to meet the needs of almost everyone in their local community during a military occupation was genuinely incredible, both in its resourcefulness and in the solidarity that it represented.

Biosynthetic analogue insulin is, of course, much more complex to produce than its animal-derived equivalents. There is no way around the fact that what the OIF is working towards is very challenging. Not only do they have to develop an effective manufacturing process, they also have to ensure that it is replicable in communities globally at manageable cost, particularly if it is going to benefit those living in serious economic deprivation.

Nonetheless, the group is entirely serious in its vision. Unlike the Saxls, its members are not amateurs. Collectively, the team has considerable expertise in biomedical research, manufacturing, and administration, and it possesses a clear passion for the mission. While it will undoubtedly be an uphill struggle, it is certainly not beyond the realms of possibility that, in time, they may achieve something quite revolutionary indeed.

Unfortunately, time is something that many of the most vulnerable do not have. They require insulin *now*, not in a few decades. Of course, all of this would be rendered irrelevant by universal access to healthcare enshrined in law. This, however, does not appear to be imminently forthcoming. In the United States, for example, mounting public attention has led to some legislative action, but most of it is extremely limited in scope.

Since 2020, insurance companies have been legally permitted to offer policies in which insulin – as a 'preventative' treatment – is not subject to the same deductible requirements as other medication, but this is wholly optional.[61] Several state governments have also now passed bills that cap 'co-pays' at $35 for the insured, and, when Joe Biden assumed the presidency in 2021, he announced that a similar provision would be included for all at the federal level through the Build Back Better Plan.

As it turned out, this piece of legislation, which had been included in 2022's Inflation Reduction Act, was – perhaps predictably – effectively shot down by Republican opposition in the Senate. In practice, however, it had only ever amounted to a nice soundbite.[62] 'Co-pay' caps do nothing to help the most desperate, uninsured Americans, who must still buy their insulin for whatever the list price is, and, even if it had covered them, $35 per month is still not exactly cheap for those already struggling to get by, especially if several different types are required.[63]

Effective political solutions are further hampered by widespread industry influence in advocacy organizations ostensibly there to act in the interests of people with diabetes. The Juvenile Diabetes Research Foundation (JDRF), for example, provides a great deal of funding towards research for a 'cure', and much of its work is undoubtedly valuable. Nonetheless, it takes vast amounts of money from the major manufacturers, creating an obvious conflict of interest. In practice, this means that it shies away from advocating for serious legislative and economic reform.[64]

Sometimes, these organizations go further. One of the most egregious examples is Beyond Type 1, well known for its celebrity co-founder Nick Jonas. Like JDRF, it is happy to take money from and work with each of the major manufacturers. This has led to its promotion of superficial initiatives like the Affordable Insulin Project – a conservative, industry-sponsored programme that goes little further than to advertise PAPs and 'co-pay' cards.[65] Beyond Type 1 is regularly embroiled in controversy for its links with the pharmaceutical industry, and on some occasions it has even worked quite plainly to block important affordability legislation. For example, in 2021 the organization wrote to the Maine legislature, which was debating an insulin safety net bill. Far from supporting the measure, however, it took the opportunity to suggest that it was no longer necessary. Instead, it promoted GetInsulin.org, an online tool constructed with industry funding that primarily works to direct individuals to manufacturer-sponsored accessibility schemes like PAPs.[66] This, Beyond Type 1 suggested, had

Figure 7.　*Patient advocates with T1International deliver insulin vials to Eli Lilly.*
Inside the vials are notes about what they have had to sacrifice due to the high cost of
insulin, 2017
Courtesy of T1International

made the legislation redundant, and by proceeding with it politicians
would 'create an administrative and financial burden to pharmacies,
patients, and the state government'.[67] It is difficult not to agree with the
sentiment – widespread amongst diabetes activists – that such organizations do more harm than good, and serve primarily to maintain the
status quo.[68] With friends like these, who needs enemies?

There is, however, room for a measure of hope. Some advocacy
organizations *are* worth paying attention to. T1International, for example, agitates for accessible insulin globally (see Figure 7). One of this
pressure group's most influential successes has been the #Insulin4all
campaign, which, despite its Twitter-influenced name, now encom-
passes a vast number of collaborators who work both on and offline to
drag the reality of the crisis into the open where it cannot be ignored.
The importance of this work cannot be overestimated.

In 2020, following months of effort, Minnesota's #Insulin4all chapter successfully pushed the state-level legislature to pass landmark

accessibility legislation despite significant opposition from manufacturers. Appropriately, the bill was named in honour of Alec Smith. The law serves to establish more affordable insurance options for those with diabetes and, importantly, ensures that anyone – insured or not – is able to access an emergency thirty-day supply for only $35 should they be unable to meet the usual costs.[69] This, of course, is only a small victory, and there is a long way yet to go. Nonetheless, the work of activists and organizations like T1International has successfully kept the accessibility crisis in the news. Despite inertia at the federal level, there is evidence that the message is beginning to get through to at least some politicians.

In 2022, for example, the Californian state legislature approved a budget that openly challenged the dominance of the major manufacturers by allocating $100.7 million – and a further $700,000 each year until at least 2026 – to domestic insulin production. Half of this money will be used for the research and development of cheaper analogues, with the rest going towards a dedicated state manufacturing facility.[70] It remains to be seen how this bold plan will play out in the long term, but, if California stays the course, it will undoubtedly be a step in the right direction.

As an author, and as someone with T1DM, I can only applaud the work of T1International and the many other groups and individuals aligned with them, and I hope that their example encourages others to take a stand. Their fight is, and will continue to be, a difficult one, but it is also vitally necessary. The global insulin crisis is not down to supply problems. It is not a necessary sacrifice for 'innovation'. It is not even the product of some abstract notion of 'greed'. It is everything to do with capitalism as an economic system. Without addressing this reality in its full scope, it is difficult to see how any satisfactory long-term resolution can be achieved.

In 2021, Eli Lilly's CEO David Ricks 'earned' $23.7 million even as thousands died for lack of insulin.[71] The problem, however, goes beyond any one pharmaceutical product. As COVID-19 swept the planet, Pfizer, which owns the patent rights to BioNTech's Comirnaty mRNA vaccine, did very well for itself indeed, doubling its profits to over $80 billion on the back of the pandemic.[72] CEO Albert Bourla personally made over $20 million in 2020.[73]

In April 2021, Tedros Adhanom Ghebreyesus, the Director-General of the WHO, drew attention to the 'shocking imbalance' between rich and poor countries when it came to COVID-19 vaccine supplies.[74] While the wealthier corners of the globe have very high vaccination rates and have successfully taken much of the sting out of the pandemic, it continues to rage unabated elsewhere, while those few international programmes designed to promote global access, such as UNICEF's COVAX, have amounted to a drop in the ocean.

Bourla, of course, claimed that Ghebreyesus was speaking 'emotionally'.[75] In 2022, it seems clear that Banting was right on at least one thing when he fought Collip in that laboratory a century ago. When it comes to those who would profiteer from the desperation of others, standing by and doing nothing cannot be justified. If the pharmaceutical industry will not voluntarily do what is necessary, then perhaps we should not shy away from taking a leaf out of his book.

Notes

Preface

1 Peter Corris, *Sweet & Sour: A Diabetic Life* (Lismore, NSW: Southern Cross University, 2000).

2 Juliet Corbin and Anselm L. Strauss, 'Accompaniments of chronic illness: change in body, self, biography, and biographical time', in Julius A. Roth and Peter Conrad (eds), *The Experience and Management of Chronic Illness* (Greenwich: JAI Press, 1987), pp. 249–81.

Introduction: What Is Insulin and Why Does It Matter?

1 Wilford Watkins-Pitchford, 'A case of rapidly fatal diabetes mellitus in a boy aged 10', *BMJ* 1639 (1892), pp. 1136–7. Until 1980, the *BMJ* identified its publications only by date. Unlike some other journals, it has not retroactively allocated volume numbers to its older material. This can sometimes make it quite difficult to find individual articles in its archive. To simplify the process, I have chosen to use issue numbers in place of volume numbers when citing it.

2 In certain fish, insulin is made in a specialized organ known as the Brockmann body, but its function is otherwise the same.

3 DKA is very different to safe nutritional ketosis, which occurs in those without diabetes when they cut carbohydrates from their diet, and involves blood ketone levels at least ten times higher. While most

common amongst those with T1DM, ketoacidosis can also be found in cases of severe starvation, and can also be a complication of alcohol abuse. There is some limited evidence that extremely strict ketogenic diets involving radical carbohydrate restriction may involve some risk, but this is far from established. For one example see Louise von Geijer and Magnus Ekelund, 'Ketoacidosis associated with low-carbohydrate diet in a non-diabetic lactating woman: a case report', *Journal of Medical Case Reports* 9 (2015), https://doi.org/10.1186/s13256-015-0709-2.

4 Because early damage to the islet cells causes few symptoms, healthcare professionals often diagnose T1DM relatively late. In the UK, for example, a Wales-based study found that, in the 1990s and 2000s, a full quarter of diagnoses were made following an emergency hospital admission for DKA; A.J. Lansdown, 'Prevalence of ketoacidosis at diagnosis of childhood onset Type 1 diabetes in Wales from 1991 to 2009 and the effect of a publicity campaign', *Diabetic Medicine* 29 (2012), pp. 1506–9.

5 National Diabetes Data Group, 'Classification and diagnosis of diabetes mellitus and other categories of glucose intolerance', *Diabetes* 28 (1979), pp. 1041–2.

6 'Dropsy' referred to any condition involving water retention, which we now know usually involves problems with either the heart or the kidneys; C.R. Bree, 'Cases of diabetes with remarks', *Provincial Medical and Surgical Journal* 1 (1840), pp. 195–6.

7 Persistently very high blood glucose levels in T2DM can, however, lead to a condition known as hyperosmolar hyperglycaemic state (HHS). This is distinct from DKA and ketones are not usually present, but it can nonetheless constitute a medical emergency; Francisco J. Pasquel and Guillermo E. Umpierrez, 'Hyperosmolar hyperglycemic state: a historic review of the clinical presentation, diagnosis, and treatment', *Diabetes Care* 37 (2014), pp. 3124–31.

8 Chris Feudtner, *Bittersweet: Diabetes, Insulin, and the Transformation of Illness* (Chapel Hill, NC: University of North Carolina Press, 2003), p. 36.

9 There is also a distinct condition called diabetes insipidus, which involves frequent urination but is rarely particularly threatening. Where I use the term 'diabetes', however, I am always referring to diabetes mellitus.

10 This could of course refer to a great deal of considerably more mundane health complaints involving frequent urination, such as urinary tract infections; D. Lynn Loriaux, 'Diabetes and the Ebers papyrus, 1552 B.C.', *The Endocrinologist* 16 (2006), pp. 55–6.

11 L.L. Frank, 'Diabetes mellitus in the texts of old Hindu medicine (Charaka, Susruta, Vagbhata)', *American Journal of Gastroenterology* 27 (1957), pp. 76–95.

12 Hui Zhang et al., 'Study on the history of Traditional Chinese Medicine to treat diabetes', *European Journal of Integrative Medicine* 2 (2010), pp. 41–6.

13 Aretaeus of Cappadocia, quoted in Folke Henschen, 'On the term diabetes in the works of Aretaeus and Galen', *Medical History* 13 (1969), p. 190.

14 Quoted in Frank N. Allan, 'The writings of Thomas Willis, M.D.', *Diabetes* 2 (1953), p. 74.

15 Ibid., p. 76.

16 Ibid., p. 77.

17 George Harley, *Diabetes: Its Various Forms and Different Treatments* (London: Walton and Maberly, 1866), p. 33.

18 John Rollo, 'An account of two cases of the diabetes mellitus, with remarks', *Annals of Medicine (Edinburgh)* 2 (1797), pp. 85–105.

19 Emphasis in original; Daniel Noble, 'A question as to pathological distinctions in cases of diabetes', *BMJ* 107 (1863), p. 59.

20 Harley describes 'defective assimilation' – almost certainly representing his patients with T1DM – as the 'second class of cases', strongly implying that they were rare; Harley, *Diabetes*, pp. 31–4.

21 Lancereaux is also often credited as one of the first to distinguish firmly between T1DM and T2DM with his 'diabète maigre' (thin/emaciated diabetes) and 'diabète gras' (fat diabetes) respectively, on the basis of pancreatic damage in the former. More recently, however, this position has received criticism, not least because the autoimmune process involved in T1DM causes damage only to the tiny insulin-producing islet cells, which would not have been perceptible to him. Instead, it has been suggested that while 'diabète gras' probably was synonymous with T2DM, 'diabète maigre' likely referred to cases of non-autoimmune damage to the pancreas; James R. Wright and Lynn McIntyre, 'Misread and mistaken: Étienne Lancereaux's endur-

ing legacy in the classification of diabetes mellitus', *Journal of Medical Biography* 30 (2022), pp. 15–20.

22 Josef von Mering and Oskar Minkowski, 'Diabetes mellitus nach Pankreasexstirpation', *Zbl Klin Med* 10 (1889), p. 393.

23 Charles-Édouard Brown-Séquard, 'Recherches expérimentales sur la physiologie et la pathologie des capsules surrénales', *Archives Générales de Médicine* 5 (1856), pp. 385–401.

24 Charles-Édouard Brown-Séquard, 'Du role physiologique et thérapeutique d'un suc extrait de testicules d'animaux: d'après nombre de faits observés chez l'homme', *Archives de Physiologie Normale et Pathologique* 5 (1889), pp. 739–46.

25 Ernest Starling, 'The Croonian lectures: on the chemical correlation of the functions of the body', *The Lancet* 166 (1905), pp. 339–41.

26 Bert Hansen, 'New images of a new medicine: visual evidence for the widespread popularity of therapeutic discoveries in America after 1885', *Bulletin of the History of Medicine* 73 (1999), pp. 629–78; Merriley Borell, 'Brown-Séquard's organotherapy and its appearance in America at the end of the nineteenth century', *Bulletin of the History of Medicine* 50 (1976), pp. 309–20.

27 George R. Murray, 'Note on the treatment of myxœdema by hypodermic injections of an extract of the thyroid gland of a sheep', *BMJ* 1606 (1891), pp. 796–7.

28 Stefan Slater, 'The discovery of thyroid replacement therapy. Part 3: A complete transformation', *Journal of the Royal Society of Medicine* 104 (2011), pp. 100–6.

29 Charles-Édouard Brown-Séquard, 'On a new therapeutic method consisting in the use of organic liquids extracted from glands and other organs', *BMJ* 1693 (1893), pp. 1146–7.

30 Robert Tattersall, *Diabetes: The Biography* (Oxford: Oxford University Press, 2009), p. 42.

31 Robert Tattersall, 'Pancreatic organotherapy for diabetes, 1889–1921', *Medical History* 39 (1995), pp. 288–316.

32 Brown-Séquard, 'On a new therapeutic method', p. 1147.

33 William Osler, *The Principles and Practice of Medicine* (New York, NY: D. Appleton and Company, 1896), p. 323.

34 Frederick Allen et al., *Total Dietary Regulation in the Treatment of Diabetes* (New York, NY: Rockefeller Institute for Medical Research, 1919), p. 598.

194   NOTES TO PP. 12–18

35 Ibid., p. 33.

36 Ibid., p. 646.

37 Frederick Allen, 'Studies concerning diabetes', *Journal of the American Medical Association* 63 (1914), pp. 939–43.

38 Allen et al., *Total Dietary Regulation*, pp. 533–4.

39 Elliott Joslin, 'The treatment of diabetes mellitus', *Canadian Medical Association Journal* 6 (1916), p. 673.

40 Allen et al., *Total Dietary Regulation*, pp. 420–8. For reference, in one village during the Nazi occupation of Poland, Jews – a population the administration was actively trying to exterminate – received an (official) ration of 300–500 calories per day. It is likely that in practice many got far less, but this does highlight quite how severe many of Allen's dietary prescriptions actually were; Lizzie Collingham, *The Taste of War: World War Two and the Battle for Food*, 2nd edn (London: Penguin, 2012), p. 206.

41 Allen et al., *Total Dietary Regulation*, pp. 568–71.

42 Ibid., p. 579.

43 Ibid., pp. 198–204.

44 Otto Leyton, *Three Lectures on the Treatment of Diabetes Mellitus by Alimentary Rest (The 'Allen' Treatment)* (London: Adlard & Son & West Newman, 1917), p. 1.

45 George Graham, 'The Goulstonian lectures on glycæmia and glycosuria', *The Lancet* 197 (1921), pp. 1059–65.

46 Allan Mazur, 'Why were "starvation diets" promoted for diabetes in the pre-insulin period?', *Nutrition Journal* 10 (2011), https://doi.org/10.1186/1475-2891-10-23.

47 Allen et al., *Total Dietary Regulation*, p. 581.

48 G.A. Wrenshall et al., *The Story of Insulin: Forty Years of Success Against Diabetes* (London: Bodley Head, 1962), p. 87.

49 Allen et al., *Total Dietary Regulation*, p. 595.

50 Elizabeth Hughes, *Letter to Mother and Father*, 24 September 1922, Ms. Coll. 334 (Hughes) Box 1, Folder 36, Hughes (Elizabeth) Papers, Thomas Fisher Rare Book Library, University of Toronto.

51 Christopher J. Rutty, '"Couldn't live without it": diabetes, the costs of innovation and the price of insulin in Canada, 1922–1984', *Canadian Bulletin of Medical History* 25 (2008), pp. 410–11.

52 It should not need to be pointed out that this also places many

people into positions in which they feel unable to leave unhappy – and perhaps even abusive – domestic situations because they rely upon a spouse or parent's insurance. For example, @CaraThe5imian ('Cara the Sim'), 'Try being a type one diabetic stuck in an abusive relationship because you can't afford your insulin but your abuser can', *Twitter*, 14 February 2020, https://twitter.com/CaraThe5imian/status/1228291420034752513.

53 For example, Christopher Keelty, 'If "Breaking Bad" had been set in the U.K.', *Imgur*, 2013, <https://imgur.com/gallery/zCmmBnc>.

54 For example, Chelsea Ritschel, 'Parents are sharing what it cost them to give birth in America: "He's 12, I still get calls from collections"', *The Independent*, 23 March 2021, <https://www.independent.co.uk/lifestyle/birth-america-cost-us-hospital-c-section-b1765878.html>.

55 IDF, *IDF Diabetes Atlas*, 10th edn (Brussels: International Diabetes Federation, 2021), p. 4.

Chapter 1: Toronto, 1921–1923

1 Michael Bliss, *The Discovery of Insulin*, 25th Anniversary Edition (Chicago, IL: University of Chicago Press, 2007).

2 Michael Bliss, *Banting: A Biography* (Toronto, ON: Toronto University Press, 1992), pp. 15–43.

3 Frederick Banting, *The Story of the Discovery of Insulin*, January 1940, Ms. Coll. 76 (Banting) Box 1, Folders 9–13, F.G. Banting (Frederick Grant, Sir) Papers, Thomas Fisher Rare Book Library, University of Toronto, p. 14.

4 Banting had visited Edinburgh for a spell while in Britain during his period of recuperation, and had made some contacts within the city's medical community; Frederick Banting, 'The history of insulin', *Edinburgh Medical Journal* 36 (1929), p. 2.

5 Frederick Banting, *Daily Accounts*, 1920–1921, Ms. Coll. 76 (Banting), Box 26, Folder 3, F.G. Banting (Frederick Grant, Sir) Papers, Thomas Fisher Rare Book Library, University of Toronto.

6 Banting, *The Story of the Discovery of Insulin*, p. 16.

7 Ibid.

8 Moses Barron, 'The relation of the islets of Langerhans to diabetes with special reference to cases of pancreatic lithiasis', *Surgery, Gynecology and Obstetrics* 31 (1920), pp. 437–48.

9 For example, Mary Kirkbride, 'The islands of Langerhans after ligation of the pancreatic ducts', *Journal of Experimental Medicine* 15 (1912), pp. 101–5; William MacCallum, 'On the relation of the islands of Langerhans to glycosuria', *Bulletin of the Johns Hopkins Hospital* 20 (1909), pp. 265–74.

10 Barron, 'The relation of the islets of Langerhans to diabetes', p. 447.

11 Banting, *The Story of the Discovery of Insulin*, p. 17.

12 Frederick Banting, *Loose Leaf Notebook*, 1920–1921, Academy of Medicine, 123 (Banting), Folders 1–11, Academy of Medicine Collection, Thomas Fisher Rare Book Library, University of Toronto.

13 Banting, *The Story of the Discovery of Insulin*, p. 23.

14 John Macleod [1922], quoted in Lloyd G. Stevenson, 'J.J.R. Macleod: history of the researches leading to the discovery of insulin', *Bulletin of the History of Medicine* 52 (1978), pp. 295–312.

15 In his 1940 account, Banting suggests that he was equally unimpressed with Macleod: 'It was the first time I had ever seen the famous professor and I was not overpowered with either the man or his knowledge of research.' By this point the two men were irrevocably estranged from one another, so it is probably fair to take this clear posturing with a grain of salt; Banting, *The Story of the Discovery of Insulin*, p. 23.

16 Frederick Banting, *F.G. Banting: Account of the Discovery of Insulin*, September 1922, Ms. Coll. 76 (Banting), Box 37, Folder 2, F.G. Banting (Frederick Grant, Sir) Papers, Thomas Fisher Rare Book Library, University of Toronto.

17 In the first three months of 1921, Banting did not dip below a monthly income of C$150 from his private practice. In February, he made over C$500; Banting, *Daily Accounts*.

18 For example, Lloyd Stephenson, *Sir Frederick Banting* (Whitby, ON: Ryerson, 1946), p. 73.

19 Banting, *The Story of the Discovery of Insulin*, pp. 18–19.

20 In 1940, Banting wrote that the room 'had not been used for fifteen years, and contained the truck and dirt of the years'. Looking back a few years later in 1942, Best was more measured, simply stating that it was 'not satisfactory during the hot summer months'; Banting, *The Story of the Discovery of Insulin*, p. 25; Charles Best, 'Reminiscences of the researches which led to the discovery of insulin', *Canadian Medical Association Journal* 47 (1942), pp. 398–400.

21 Charles Best, 'A short essay on the importance of dogs in medical research', *The Physiologist* 17 (1974), pp. 437–40. A note following the main body of this article makes it clear that Best had been specifically asked to write in defence of animal testing, a subject of intense debate in the mid-1970s. He and Banting had conducted their initial experiments when anti-vivisectionist protests were a common sight in Toronto, so he certainly had a personal stake in the issue.

22 Banting, *The Story of the Discovery of Insulin*, p. 19.

23 Banting, *F.G. Banting: Account of the Discovery of Insulin*.

24 Ibid.; Macleod [1922], quoted in Stevenson, 'J.J.R. Macleod: history of the researches leading to the discovery of insulin', p. 304.

25 C.H. Best and D.A. Scott, 'The preparation of insulin', *Journal of Biological Chemistry* 57 (1923), pp. 709–23.

26 Banting, 'The history of insulin', pp. 6–7.

27 It is impossible to know exactly why this happened, but it could have been the result of hypoglycaemia or perhaps an allergic reaction – the extract Banting and Best were using had no consistent potency and contained many impurities.

28 Banting, *The Story of the Discovery of Insulin*, p. 36.

29 *Patient Records for Leonard Thompson*, December 1921–January 1922, Ms. Coll. 76 (Banting), Box 8B, Folder 17B, F.G. Banting (Frederick Grant, Sir) Papers, Thomas Fisher Rare Book Library, University of Toronto.

30 Walter Campbell, 'Anabasis', *Canadian Medical Association Journal* 87 (1962), pp. 1055–61.

31 Frank N. Allan, 'Diabetes before and after insulin', *Medical History* 16 (1972), p. 267.

32 Walter Campbell, interviewed by Robert Noble, *c.*1967, quoted in Bliss, *The Discovery of Insulin*, p. 112.

33 In a Nobel lecture in Stockholm several years later, Banting gave a rather charitable account of the first test, and moved past the topic quickly; Frederick Banting, 'Nobel lecture: Diabetes and insulin; *The Nobel Prize*, 15 September 1925, <https://www.nobelprize.org/prizes/medicine/1923/banting/lecture/>.

34 James Collip, 'The history of the discovery of insulin', *Northwest Medicine* 22 (1923), pp. 267–73.

35 Interviews with those who were close to Banting suggest that in later life, he began to openly acknowledge the role that Collip had played,

and admit that without him the project may have hit a brick wall; Bliss, *The Discovery of Insulin*, p. 237.

36 F.G. Banting et al., 'Pancreatic extracts in the treatment of diabetes mellitus', *Canadian Medical Association Journal* 12 (1922), p. 142.

37 F.G. Banting and C.H. Best, 'The internal secretion of the pancreas', *Journal of Laboratory and Clinical Medicine* 7 (1922), p. 265.

38 Banting et al., 'Pancreatic extracts in the treatment of diabetes mellitus', p. 146.

39 The paper was later published, appended with excerpts from the discussion; F.G. Banting et al., 'The effect produced on diabetes by extracts of pancreas', 1922, Ms. Coll. 76 (Banting), Box 62, Folder 17, F.G. Banting (Frederick Grant, Sir) Papers, Thomas Fisher Rare Book Library, University of Toronto.

40 Elliott Joslin, 'Pancreatic extract in the treatment of diabetes', *Boston Medical and Surgical Journal* 186 (1922), p. 654.

41 Banting et al., 'The effect produced on diabetes by extracts of pancreas', p. 10.

42 *Patient Records for Leonard Thompson*, December 1921–January 1922, Ms. Coll. 76 (Banting), Box 8B, Folder 17B, F.G. Banting (Frederick Grant, Sir) Papers, Thomas Fisher Rare Book Library, University of Toronto.

43 Allan, 'Diabetes before and after insulin', p. 267.

44 While writing *The Discovery of Insulin* in 1980, Bliss contacted William Gossett, Hughes' husband. Expecting him to be a widower, he was amazed to receive a reply from Hughes herself, horrified that he had been able to find her at all; Bliss, *The Discovery of Insulin*, pp. 243–4.

45 The use of 'expendable' bodies in medical research does, of course, have a long history, and Thompson was fortunate both that his family were at least asked for consent and that he did benefit in the long term. This has been far from the case for many 'participants', particularly amongst non-White people. For one particularly egregious example, see Susan M. Reverby, *Examining Tuskegee: The Infamous Syphilis Study and Its Legacy* (Chapel Hill, NC: University of North Carolina Press, 2009). For more on human medical experimentation more generally, see Jordan Goodman, Anthony McElligott, and Lara Marks (eds), *Useful Bodies: Humans in the Service of Medical Science in the Twentieth Century* (Baltimore, MD: Johns Hopkins University Press, 2003).

46 Bliss, *The Discovery of Insulin*, p. 243.

47 Ibid., pp. 129–34.

48 John Macleod, *Statement read by J.J.R. Macleod at the Insulin Committee Meeting Regarding Patents and Royalties*, 28 April 1924, A1982-0001, Box 6, Folder 1, University of Toronto Board of Governors Insulin Committee Archive, Archives and Records Management Services, University of Toronto.

49 *Assignment to the Governors of the University of Toronto*, 15 January 1923, Ms. Coll. 76 (Banting), Box 10, Folder 10, F.G. Banting (Frederick Grant, Sir) Papers, Thomas Fisher Rare Book Library, University of Toronto.

50 *Indenture Between the Governors of the University of Toronto and the Eli Lilly Company*, 30 May 1922, Ms. Coll. 269 (Collip), Box 37, Folder 5, Collip (James Bertram) Papers, Thomas Fisher Rare Book Library, University of Toronto.

51 J.K. Lilly, *Letter to Dr. Clowes*, 3 January 1923, A1982-0001, Box 12, Eli Lilly Folder, University of Toronto Board of Governors Insulin Committee Archive, Archives and Records Management Services, University of Toronto.

52 For more on Lilly's early commercial strategy, see John Patrick Swann, 'Insulin: a case study in the emergence of collaborative pharmacomedical research', *Pharmacy in History* 28 (1986), pp. 65–74.

53 John Macleod, *Letter to Sir Walter Fletcher*, 17 January 1923, A1982-0001, Box 15, University of Toronto Board of Governors Insulin Committee Archive, Archives and Records Management Services, University of Toronto.

54 Maurice Cassier and Christiane Sinding, '"Patenting for the public interest": administration of insulin patents by the University of Toronto', *History and Technology* 24 (2008), pp. 153–71.

55 Banting, *The Story of the Discovery of Insulin*, pp. 63–4; Frederick Banting, *Letter Accepting an Appointment in the Diabetic Clinic in the Toronto General Hospital*, 17 July 1922, A1967-0007, Box 81, Office of the President Archive, Archives and Records Management Services, University of Toronto.

56 House of Commons (Canada), *Debates*, 27 June 1923, pp. 4437–40.

57 These were likely the result of hypoglycaemia, because the potency of Zülzer's extract was highly variable; Georg Zülzer, 'Experimentell

Untersuchungen über den Diabetes', *Berliner Klinische Wochenschrift* 44 (1907), pp. 474–5; Georg Zülzer, 'Ueber Versuche einer specifischen Fermenttherapie des Diabetes', *Zeitschrift für Experimentelle Pathologie und Therapie* 5 (1908), pp. 307–18.

58 Banting, 'The history of insulin', pp. 7–8.

59 Lydia DeWitt, 'Morphology and physiology of areas of Langerhans in some vertebrates', *Journal of Experimental Medicine* 8 (1906), pp. 193–239.

60 Eugène Gley, 'Action des extraits de pancréas sclérosé sur des chiens diabétiques (par extirpation du pancréas)', *Comptes Rendus des Séances de la Société de Biologie* 87 (1922), pp. 1322–5.

61 Bliss, *The Discovery of Insulin*, pp. 208–9.

62 In their first article in February 1922, Banting and Best state that the whole pancreas extracts are 'much weaker than that from the degenerated gland'. This was simply not true according to the results they recorded; Banting and Best, 'The internal secretion of the pancreas', p. 256.

63 Israel S. Kleiner, 'The action of intravenous injections of pancreas emulsions in experimental diabetes', *Journal of Biological Chemistry* 40 (1919), p. 170.

64 Robert Tattersall, *The Pissing Evil: A Comprehensive History of Diabetes Mellitus* (Fife: Swan & Horn, 2017), p. 85.

65 Ernest Scott, 'The effect of pancreas extract on pancreatized dogs' (unpublished MS Thesis, University of Chicago, 1911), pp. 8–9.

66 Scott's work was published on his behalf, albeit with slightly moderated conclusions. Most importantly, a passage was inserted stating that 'It does not follow that these effects are due to the internal secretion of the pancreas in the extract'; Ernest Scott, 'On the influence of intravenous injections of an extract of the pancreas on experimental pancreatic diabetes', *American Journal of Physiology* 29 (1912), p. 310.

67 Banting, *The Story of the Discovery of Insulin*, p. 14; Ernest Lyman Scott, *Letter to F.G. Banting*, 23 November 1923, Ms. Coll. 76 (Banting) Box 1, Folder 38, F.G. Banting (Frederick Grant, Sir) Papers, Thomas Fisher Rare Book Library, University of Toronto. After his death in 1966, Scott's widow Aleita took it upon herself to defend his legacy. Her self-published book is a bizarre document and had little impact, but it makes her feelings towards the official 'discoverers' abundantly

clear; Aleita Scott, *Great Scott: Ernest Lyman Scott's Work with Insulin in 1911* (Bogota, NJ: Scott Publishing, 1972).

68 Nicolae Paulescu, 'Action de l'extrait pancréatique injecté dans le sang, chez un animal diabétique', *Comptes rendus hebdomadaires des séances et mémoires de la Société de Biologie et de ses filiales* 85 (1921), pp. 555–9; Nicolae Paulescu, 'Quelques réactions chimiques et physiques, appliquée à l'extrait aqueux du pancréas, des substances protéiques en excès', *Archives Internationales de Physiologie* 21 (1923), pp. 71–85.

69 Following the publication of a 1971 article that claimed precedence for Paulescu, the controversy was reignited. It still remains a source of bitterness for some Romanian scholars; Ian Murray, 'Paulesco and the isolation of insulin', *Journal of the History of Medicine* 26 (1971), pp. 150–7.

70 Banting does not appear to have noticed a vicious critique published in the *BMJ* at the end of 1922, which described the initial experiments as 'wrongly conceived, wrongly conducted, and wrongly interpreted'; Ffrangcon Roberts, 'Insulin', *BMJ* 3233 (1922), pp. 1193–4.

71 The quote is from Henry Dale's response to Roberts' critique; Henry Dale, 'Insulin', *BMJ* 3234 (1922), p. 1241.

72 Compiling his account of the 'discovery' in 1922, Collip directly referenced Scott, and clearly recognized the quality of his work. He wrote that 'Scott came so near to obtaining the active principle that it is hard to understand why he did not pursue his work further.' It is unclear whether Collip was being deliberately wry, or if he was simply naïve; Collip, 'The history of the discovery of insulin', p. 9.

Chapter 2: Insulin in Practice, 1922–1978

1 The headlines quoted here are taken from clippings in Banting's scrapbooks, held by the University of Toronto's Thomas Fisher Rare Book Library.

2 'Now hoped that insulin will be permanent cure', *Toronto Star*, *c.* January 1923, Ms. Coll. 76 (Banting) Scrapbook 1, Boxes I–III, F.G. Banting (Frederick Grant, Sir) Papers, Thomas Fisher Rare Book Library, University of Toronto.

3 F.G. Banting et al., 'Insulin in the treatment of diabetes mellitus', *Journal of Metabolic Research* 2 (1922), p. 599. This journal is listed as the November–December 1922 edition, but publication was delayed

until May 1923. This led to the strange situation of articles in a journal nominally dated 1922 referencing material from early 1923.

4 William McCann, quoted in 'Insulin is not diabetes cure, says physician', *Unattributed Newspaper Clipping*, July 1923, Ms. Coll. 76 (Banting) Scrapbook 1, Boxes I–III, F.G. Banting (Frederick Grant, Sir) Papers, Thomas Fisher Rare Book Library, University of Toronto.

5 'The cure of diabetes', *Westminster Gazette*, 20 April 1923, p. 6.

6 Collip, 'History of the discovery of insulin', pp. 7–8. For one outline of the process, see Melville Sahyun and N.R. Blatherwick, 'The rabbit method of standardizing insulin', *American Journal of Physiology* 76 (1926), pp. 677–84. For more on the topic of insulin's standardization, see Christiane Sinding, 'Making the unit of insulin: standards, clinical work, and industry, 1920–1925', *Bulletin of the History of Medicine* 76 (2002), pp. 231–70; Desirée Cox-Maksimov, 'The making of the clinical trial in Britain, 1910–1945: expertise, the state, and the public' (unpublished PhD thesis, University of Cambridge, 1997).

7 F.G. Banting et al., 'Further clinical experience with insulin (pancreatic extracts) in the treatment of diabetes mellitus', *BMJ* 3236 (1923), pp. 9–10.

8 This could be made from either beef or pork pancreases, but in practice both types worked similarly. Some researchers also attempted to produce insulin from other animals. Macleod, for example, had some success with fish, but dropped the idea on grounds of practicality. During the Second World War, however, both fish and whale insulin were used extensively in Japan; John Macleod, 'Pancreatic extract and diabetes', *Canadian Medical Association Journal* 12 (1922), pp. 423–5; Kakuma Nagasawa, 'Use of fish and whale insulin as drugs in Japan', *Journal of Association of Official Analytical Chemists* 51 (1968), pp. 326–9.

9 As a European, I use soluble by default.

10 William McCann, quoted in 'Insulin is not diabetes cure, says physician'.

11 Frederick Allen quoted in 'Diabetics have to diet or insulin powerless', *Toronto Star*, November 1923, Ms. Coll. 76 (Banting) Scrapbook 1, Boxes I–III, F.G. Banting (Frederick Grant, Sir) Papers, Thomas Fisher Rare Book Library, University of Toronto.

12 'Insulin is deadly in amateur's hands', *Toronto Star*, c. June 1923, Ms. Coll. 76 (Banting) Scrapbook 1, Boxes I–III, F.G. Banting (Frederick

Grant, Sir) Papers, Thomas Fisher Rare Book Library, University of Toronto.

13 Elliott Joslin, *A Diabetic Manual for the Mutual Use of Doctor and Patient* (Philadelphia, PA: Lea & Febiger, 1934).

14 Chris Feudtner, 'The want of control: ideas, innovations, and ideals in the modern management of diabetes mellitus', *Bulletin of the History of Medicine* 69 (1995), p. 75.

15 R.D. Lawrence, 'The beginning of the Diabetic Association in England', *Diabetes* 1 (1952), pp. 420–1.

16 Jane Lawrence and Robert Tattersall, *Diabetes, Insulin and the Life of RD Lawrence* (London: Royal Society of Medicine Press, 2012), pp. 1–19.

17 R.D. Lawrence, *The Diabetic Life*, 1st edn (London: J. & A. Churchill, 1925), p. 126.

18 Ibid., p. 115. The line 'very often the result of a holiday is the opposite of beneficial' appears in every single one of *The Diabetic Life*'s seventeen editions.

19 Lawrence, *The Diabetic Life*, 1st edn, p. 115. This line was not removed until the 14th edition in 1950, replaced with a moderated call to 'think more of himself and less of . . . [the host's] feelings'; R.D. Lawrence, *The Diabetic Life*, 14th edn (London: J. & A. Churchill, 1950), p. 169.

20 For more on Lawrence and *The Diabetic Life*, see Martin D. Moore, 'Food as medicine: diet, diabetes management, and the patient in twentieth century Britain', *Journal of the History of Medicine and Allied Sciences* 73 (2018), pp. 150–67.

21 For example, Mo Linton, *Diabetes Stories*, <http://diabetes-stories .com/transcript.asp?UID=73>.

22 Lawrence, *The Diabetic Life*, 1st edn, pp. 126–7.

23 Elliott Joslin, 'The routine treatment of diabetes with insulin', *Journal of the American Medical Association* 80 (1923), p. 1581.

24 Lawrence, *The Diabetic Life*, 1st edn, p. 126.

25 For example, Monica Winn, *Diabetes Stories*, <http://www.diabetes -stories.com/interview.asp?UID=31>.

26 'Grace', *Diabetes Stories*, <http://www.diabetes-stories.com/intervi ew.asp?UID=13>.

27 Gillian Clifton interviewed by author, 8 July 2017, GB 249 SOHC 64, University of Strathclyde Archives and Special Collections.

28 Carol Cowan interviewed by author, 20 June 2017, GB 249 SOHC 64, University of Strathclyde Archives and Special Collections.

29 Robert Freudenthal and Joanna Moncrieff, '"A landmark in psychiatric progress"? The role of evidence in the rise and fall of insulin coma therapy', *History of Psychiatry* 33 (2022), pp. 65–78; Benjamin Zajicek, 'Insulin coma therapy and the construction of therapeutic effectiveness in Stalin's Soviet Union, 1936–1953', in Mat Savelli and Sarah Marks (eds), *Psychiatry in Communist Europe* (London: Palgrave Macmillan, 2015), pp. 50–72; S.M. Hillier, 'Psychiatry and the treatment of mental illness in China', in S.M. Hillier and J.A. Jewell (eds), *Health Care and Traditional Medicine in China, 1800–1982* (London: Routledge, 1983), pp. 378–407.

30 Ron Howard (director), *A Beautiful Mind* (2001), Universal Pictures.

31 Initially, most ICT units followed the guidelines established by Sakel, but over time psychiatrists began to use their own experience to tailor their practice; Manfred Sakel, *The Classical Sakel Shock Treatment, a Reappraisal: The Basic Principles of the Sakel Treatment* (New York, NY: Manfred Sakel Foundation, 1954).

32 Deborah Blythe Doroshow, 'Performing a cure for schizophrenia: insulin coma therapy on the wards', *Journal of the History of Medicine and Allied Sciences* 62 (2007), p. 236.

33 One 2014 article estimated a mortality rate of around 1%, but suggested that 'it is highly likely that, because of the upbeat enthusiasm with which the therapy ... was embraced, any negative headlines were brushed aside and largely ignored'; Jonathan Pimm, 'Profile: Dr Bourne's identity – credit where credit's due', *Psychiatric Bulletin* 38 (2014), p. 83.

34 Harold Bourne, 'The insulin myth', *The Lancet* 262 (1953), pp. 964–8; Harold Bourne, cited in Doroshow, 'Performing a cure for schizophrenia', p. 243.

35 Alan Gibson, 'Insulin coma therapy', *Psychiatric Bulletin* 38 (2014), p. 198.

36 Arthur Cain, cited in Doroshow, 'Performing a cure for schizophrenia', p. 238.

37 Don Weitz, 'Notes of a "schizophrenic" shitdisturber', in Bonnie Burstow and Don Weitz (eds), *Shrink Resistant: The Struggle Against Psychiatry in Canada* (Vancouver, BC: New Star Books, 1988), p. 287.

38 These thinkers are often lumped together as members of the so-called 'antipsychiatry' movement, but this oversimplification is not particularly helpful, and many of them rejected the label.

39 Cheryl McGeachan, 'Needles, picks and an intern named Laing: exploring the psychiatric spaces of Army life', *Journal of Historical Geography* 40 (2013), pp. 67–78.

40 Lawrence, *The Diabetic Life*, 1st edn, p. 127.

41 'Peter', *Diabetes Stories*, <http://www.diabetes-stories.com/intervi ew.asp?UID=42>.

42 Lawrence, *The Diabetic Life*, 1st edn, p. 88.

43 R.D. Lawrence, *The Diabetic Life*, 17th edn (London: J. & A. Churchill, 1965), p. 202.

44 Peter Davies interviewed by author, 10 May 2017, GB 249 SOHC 64, University of Strathclyde Archives and Special Collections.

45 Clifton, Interview.

46 John Meredith interviewed by author, 7 July 2017, GB 249 SOHC 64, University of Strathclyde Archives and Special Collections.

47 There were some attempts to create a model standard, particularly for the benefit of those non-specialist 'busy practitioners' who did not have the time to 'constantly ... work out different diets', but in practice few prescribed diets were exactly alike; for example, A. Clarke Begg, 'A standard diet for insulin treatment', *The Lancet* 203 (1924), pp. 516–17.

48 'Reference intakes explained', *NHS* <https://www.nhs.uk/live-we ll/eat-well/what-are-reference-intakes-on-food-labels/>. Begg's above cited 'standard diet' was comprised of seventy grams of carbohydrate, eighty grams of protein, and 185 (!) grams of fat daily.

49 This is explicitly reflected in initiatives like the Joslin Diabetes Centre's medallist programme. From 1948, these were awarded at twenty-five years post-diagnosis. Tellingly, only those without complications were eligible. Today, this stipulation is no longer applied, and a quarter-century nets the recipient only a certificate. Medals continue to be given to those surviving fifty, seventy-five and eighty years. Other organizations like Diabetes UK also have similar initiatives; Feudtner, *Bittersweet*, pp. 188–90.

50 'The price of insulin', *The Lancet* 203 (1924), p. 346.

51 Rutty, '"Couldn't live without it"', pp. 410–11.

52 Robert Tattersall; 'A force of magical activity: the introduction of insulin treatment in Britain, 1922–1926', *Diabetic Medicine* 12 (1995), p. 745.

53 Caleb Williams Seleeby, 'The triumph of "insulin"', *Daily Chronicle*, 17 May 1923.

54 Colin Clark, *National Income and Outlay* (London: Frank Cass & Co., 1965), p. 69.

55 C.J.C. Earl, 'The treatment of diabetics as hospital out-patients', *BMJ* 3461 (1927), pp. 831–3.

56 *Recipes and Diet Sheets, c.*1925–1927, item 19, MS5939/2, Alan Nabarro Collection, Royal College of Physicians, London.

57 *Geoffrey Harrison to Alan Nabarro*, 21 October 1927, item 20, MS5941/1, Alan Nabarro Collection, Royal College of Physicians, London.

58 *Bar Mitzvah Diet Sheet*, November 1927, item 21, MS5939/2, Alan Nabarro Collection, Royal College of Physicians, London.

59 Lawrence, *The Diabetic Life*, 1st edn, pp. 33–4.

60 'Ann', *Diabetes Stories*, <http://diabetes-stories.com/transcript.asp?UID=1>.

61 Clifton, Interview.

62 Lis Warren interviewed by author, 5 May 2017, GB 249 SOHC 64, University of Strathclyde Archives and Special Collections.

63 Joanne Pinfield, *Diabetes Stories*, <http://diabetes-stories.com/transcript.asp?UID=30>.

64 Clifton, Interview.

65 Lawrence, *The Diabetic Life*, 17th edn, p. 158.

66 For example, Lee Ducat et al., 'The mental health comorbidities of diabetes', *Journal of the American Medical Association* 312 (2014), pp. 691–2.

67 Ingrid Torjesen, 'Diabulimia: the world's most dangerous eating disorder', *BMJ* 8190 (2019), https://doi.org/10.1136/bmj.l982.

68 Sophie Elizabeth Coleman and Noreen Caswell, 'Diabetes and eating disorders: an exploration of "diabulimia"', *BMC Psychology* 8 (2020), https://doi.org/10.1186/s40359-020-00468-4.

69 Some, tragically, have taken their own lives as a result; for example, Tracy Ollerenshaw, 'The suicide note that told Megan's diabulimia story', *BBC*, 25 September 2017, <https://www.bbc.co.uk/news/newsbeat-40888659>.

70 R.D. Lawrence, *The Diabetic Life*, 12th edn (London: J. & A. Churchill, 1941), p. 80.

71 R.D. Lawrence, *The Diabetic Life*, 15th edn (London: J. & A. Churchill, 1955), pp. 142–5.

72 Ibid., p. 145.

73 Emphasis in original; ibid., p. 87.

74 Lawrence, *The Diabetic Life*, 15th edn, pp. 87, 145.

75 H.P. Himsworth, 'High carbohydrate diets and insulin efficiency', *BMJ* 3836 (1934), p. 58.

76 Ibid., p. 60.

77 This interest in insulin sensitivity led Himsworth to speculate 'of the existence of a type of diabetes due not to diminished secretion of the insulin by the pancreas, but to a greater or less impairment of the organism's susceptibility to insulin', clearly describing T2DM. Later, he became one of the first to make a firm distinction between the two major types of diabetes in his practice; H.P. Himsworth, 'Diabetes mellitus: the differentiation into insulin-sensitive and insulin-insensitive types', *The Lancet* 227 (1936), pp. 127–30.

78 While the professions of Lewis' parents are not recorded, he and his family lived in the Queen's Buildings, a Victorian-era working-class housing development in Southwark occupied primarily by manual labourers; *Diet Sheet: Charles Lewis*, 29 December 1939, EO/WAR/2/32, London Metropolitan Archives.

79 H.C. Hagedorn et al., 'Protamine insulinate', *Journal of the American Medical Association* 106 (1936), pp. 177–80.

80 Emphasis in original; R.D. Lawrence, 'Zinc-protamine-insulin in diabetes: treatment by one daily injection', *BMJ* 4090 (1939), p. 1079.

81 Edward Tolstoi, 'Newer concepts in the treatment of diabetes mellitus with protamine insulin', *American Journal of Digestive Diseases* 10 (1943), p. 248.

82 Ibid., pp. 247–8.

83 Ibid., p. 248.

84 Edward Tolstoi et al., 'Treatment of diabetes mellitus with protamine insulin: is a persistent glycosuria harmful? A metabolic study of a severe case', *Annals of Internal Medicine* 16 (1942), p. 893.

85 Ibid., p. 903.

86 'Conferences on therapy: management of diabetic emergencies: 1. General treatment', *Journal of the American Medical Association* 115 (1940), p. 454.

87 Elliott Joslin et al., 'Treatment of diabetes', *Journal of the American Medical Association* 115 (1940), p. 1039.

88 Edward Tolstoi, 'The free diet for diabetic patients', *American Journal of Nursing* 50 (1950), p. 654.

89 Edward Tolstoi, 'Treatment of diabetes with "free diet" during the past ten years', in Samuel Soskin (ed.), *Progress in Clinical Endocrinology* (London: William Heinemann, 1950), pp. 292–302.

90 C.C. Forsyth et al., 'Diet in diabetes', *BMJ* 4715 (1951), p. 1101.

91 D.M. Dunlop, 'Are diabetic degenerative complications preventable?', *BMJ* 4884 (1954), p. 384.

92 Ibid., pp. 384–5.

93 Tolstoi, 'The free diet for diabetic patients', p. 652.

94 Ibid., p. 653.

95 The terms T1DM and T2DM were, of course, not in use at the time of production. Instead, the film uses language that would be considered unacceptable today, referring to 'skinny' and 'fat' diabetes respectively; R.T. Goodcliffe (director), *Living with Diabetes* (1959), Wellcome Foundation.

96 Wrenshall et al., *The Story of Insulin*, p. 117.

97 Ruth E. Reuting, 'Progress notes on fifty diabetic patients followed twenty-five or more years', *Archives of Internal Medicine* 86 (1950), pp. 891–97.

98 For example, Hans Christian Hagedorn, 'Cases of diabetes of long duration', *New England Journal of Medicine* 261 (1959), pp. 442–3; W. Korp and Ernst Zweymüller, '50 years of insulin treatment at the Vienna Hospital for Children – the fate of diabetic children from the first insulin era' [1973], in Dietrich von Engelhardt (ed.), *Diabetes: Its Medical and Cultural History* (Berlin: Springer-Verlag, 1989), pp. 437–50.

99 Frederick Allen, 'Methods and results of diabetic treatment', *New England Journal of Medicine* 203 (1930), pp. 1133–9.

Chapter 3: 'Intensification', 1976–1993

1 For one broad review of the contemporary arguments, see G. Tchobroutsky, 'Relation of diabetic control to development of microvascular complications', *Diabetologia* 15 (1978), pp. 143–52.

2 George Cahill et al., '"Control" and diabetes', *New England Journal of Medicine* 294 (1976), p. 1004.

3 R.D. Lawrence, 'Insulin therapy: successes and problems', *The Lancet* 254 (1949), pp. 401–5.

4 W.G. Oakley et al., *Diabetes and Its Management*, 3rd edn (Oxford: Blackwell, 1978), p. 77.

5 Cahill et al., '"Control" and diabetes', p. 1004. The ADA's position did not, however, go wholly unchallenged. For example, see M.D. Siperstein et al., 'Control of blood glucose and diabetic vascular disease', *New England Journal of Medicine* 296 (1977), pp. 1060–3.

6 This differs between individuals, but is usually around 10 mmol/l – already a slightly elevated level; Steven L. Cowart and Max E. Stachura, 'Glycosuria', in H. Kenneth Walker, W. Dallas-Hall and J. Willis Hurst (eds), *Clinical Methods*, 3rd edn (Boston, MA: Butterworths, 1990), p. 653.

7 These shortcomings were often not lost on the people doing the tests. As 'Deborah' put it during an interview, many understood that 'obviously, being urine, it's kind of too late', and she was far from the only person to make similar comments; 'Deborah' interviewed by author, 13 June 2017, GB 249 SOHC 64, University of Strathclyde Archives and Special Collections.

8 S.F. Clarke and J.R. Foster, 'A history of blood glucose meters and their role in self-monitoring of diabetes mellitus', *British Journal of Biomedical Science* 69 (2012), pp. 83–93.

9 Tattersall, *Diabetes: The Biography*, p. 162.

10 Clara Lowy, 'Diabetes in pregnancy', *Journal of the Royal Society of Medicine* 71 (1978), pp. 541–2.

11 Clara Lowy, 'A memorable patient: home glucose monitoring, who started it?', *BMJ* 7142 (1998), p. 1467.

12 Tattersall, *The Pissing Evil*, pp. 270–1.

13 P.H. Sönksen et al., 'Home monitoring of blood-glucose: method for improving diabetic control', *The Lancet* 311 (1978), pp. 729–32; S. Walford et al., 'Self-monitoring of blood-glucose: improvement of

210   NOTES TO PP. 77–80

diabetic control', *The Lancet* 311 (1978), pp. 732–5; J.S. Skyler et al., 'Home blood glucose monitoring as an aid in diabetic management', *Diabetes Care* 1 (1978), pp. 150–7.

14 Tattersall, *The Pissing Evil*, p. 271.

15 Sheldon J. Bleicher, 'Chairman's introduction', *Diabetes Care* 3 (1980), p. 57.

16 Colin Dexter, *Diabetes Stories*, <http://diabetes-stories.com/transcript.asp?UID=50>.

17 Clifton, Interview.

18 For example, Judith M. Steel and Margaret Dunn, *Coping with Life on Insulin* (Edinburgh: W&R Chambers, 1987), pp. 42–3.

19 For example, R.S. Mazze et al., 'Reliability of blood glucose monitoring by patients with diabetes mellitus', *American Journal of Medicine* 77 (1984), pp. 211–17; P.L. Hoskins et al., 'Comparison of different models of diabetic care on compliance with self-monitoring of blood glucose by memory glucometer', *Diabetes Care* 11 (1988), pp. 719–24; C.D. Williams et al., 'Use of memory meters to measure reliability of self blood glucose monitoring', *Diabetic Medicine* 5 (1988), pp. 459–62; O. Ziegler et al., 'Reliability of self-monitoring of blood glucose by CSII-treated patients with Type 1 diabetes', *Diabetes Care* 12 (1989), pp. 184–8.

20 David Armstrong, 'The social context of technology in diabetes care: "compliance" and "control"', in Clare Bradley, Philip Home, and Margaret Christie (eds), *The Technology of Diabetes Care: Converging Medical and Psychosocial Perspectives* (Chur: Harwood, 1991), pp. 17–23.

21 R.J. Koenig et al., 'Correlation of glucose regulation and haemoglobin A1c in diabetes mellitus', *New England Journal of Medicine* 295 (1976), pp. 417–20.

22 Christiane Sinding, 'Flexible norms? From patients' values to physicians' standards', in Ernst Waltraud (ed.), *Histories of the Normal and the Abnormal: Social and Cultural Histories of Norms and Normativity* (Abingdon: Routledge, 2006), pp. 237–9.

23 Clifton, Interview.

24 HbA1c values amongst the general population tend to be around 5–6%, reflecting average blood sugar levels of 5–7 mmol/l. A reading of 12% would indicate persistent values of around 16.5 mmol/l; Charles Fox

interviewed by author, 24 May 2017, GB 249 SOHC 64, University of Strathclyde Archives and Special Collections.

25 A.H. Kadish, 'A servomechanism for blood sugar control', *Biomedical Sciences Instrumentation* 1 (1963), pp. 171–6; A.H. Kadish, 'Automation control of blood sugar: a servomechanism for glucose monitoring and control', *Transactions of the American Society for Artificial Internal Organs* 9 (1963), pp. 363–7; A.H. Kadish, 'Automation control of blood sugar: a servomechanism for glucose monitoring and control', *American Journal of Medical Electronics* 3 (1964), pp. 82–6.

26 E.F. Pfeiffer et al., 'The artificial beta cell – a continuous control of blood sugar by external regulation of insulin infusion (glucose controlled insulin infusion system)', *Hormone and Metabolic Research* 487 (1974), pp. 339–42; E.J. Fogt et al., 'Development and evaluation of a glucose analyser for a glucose controlled insulin infusion system (Biostator)', *Clinical Chemistry* 24 (1978), pp. 1366–72.

27 'BDA is to buy artificial pancreas', *Balance*, August 1977, pp. 1, 16.

28 F.M. Alsaleh, 'Insulin pumps: from inception to the present and toward the future', *Journal of Clinical Pharmacy and Therapeutics* 35 (2010), pp. 128–9.

29 G. Slama et al., 'One to five days of continuous intravenous insulin infusion on seven diabetic patients', *Diabetes* 23 (1974), pp. 732–8.

30 J.C. Pickup et al., 'Continuous subcutaneous insulin infusion: an approach to achieving normoglycaemia', *BMJ* 6107 (1978), pp. 204–7. The Mill Hill infuser was named after the district of London that housed Britain's National Institute for Medical Research, where it had originally been developed for studies involving parathyroid hormone.

31 Pickup reflected on this in a later commentary on the paper; John Pickup, 'Citation classic: Continuous subcutaneous insulin infusion', *Current Contents* 34 (1987), p. 14.

32 For example, William V. Tamborlane et al., 'Reduction to normal of plasma glucose in juvenile diabetes by subcutaneous administration of insulin with a portable infusion pump', *New England Journal of Medicine* 300 (1979), pp. 573–8.

33 Walford et al., 'Self-monitoring of blood-glucose', p. 735.

34 Sönksen et al., 'Home monitoring of blood-glucose', pp. 729–32.

35 For example, C.M. Peterson et al., 'Feasibility of improved blood glucose control in patients with insulin-dependent diabetes mellitus', *Diabetes Care* 2 (1979), pp. 329–35; Jay S. Skyler et al., 'Instructing patients in making alterations in insulin dosage', *Diabetes Care* 2 (1979), pp. 39–45; C.M. Peterson et al., 'Self-management: an approach to patients with insulin-dependent diabetes mellitus', *Diabetes Care* 3 (1980), pp. 82–7; Jay S. Skyler et al., 'Algorithms for adjustment of insulin dosage by patients who monitor blood glucose', *Practical Diabetes* 4 (1981), pp. 311–18.

36 Knight had been a long-time advocate of blood glucose testing in the home. In 1962, for example, he had suggested that people with diabetes might take regular samples throughout the day to post to hospital for analysis; Harry Keen and Anthony Knight, 'Self-sampling for blood-sugar', *The Lancet* 279 (1962), pp. 1037–40.

37 M. Warner, 'Diagnostic kits and the clinical chemist', *BMJ* 6204 (1979), pp. 1581–2.

38 A.H. Knight, 'Monitoring blood glucose', *BMJ* 6209 (1980), p. 253.

39 Anthony Knight, 'Blood glucose self-monitoring', *Balance*, October 1981, p. 5

40 Ibid., p. 11.

41 Knight, 'Blood glucose self-monitoring', p. 5.

42 Steel and Dunn, *Coping with Life on Insulin*, p. 60.

43 Peterson et al., 'Self-management', p. 86.

44 Philip Newick, *Diabetes Stories*, <http://diabetes-stories.com/transcript.asp?UID=39>.

45 R. Worth et al., 'Intensive attention improves glycaemic control in insulin-dependent diabetes without further advantage from home blood glucose monitoring: results of a controlled trial', *BMJ (Clinical Research Edition)* 6350 (1982), p. 1233.

46 Lawrence, 'Insulin therapy: successes and problems', p. 404.

47 For example, D.P. Walters et al., 'Experience with NovoPen, an injection device using cartridged insulin, for diabetic patients', *Diabetic Medicine* 2 (1986), pp. 496–7; T. Jensen et al., 'Metabolic control and patient acceptability of multiple insulin injections using NovoPen cartridge-packed insulin', *Practical Diabetes* 3 (1986), pp. 302–6; Ewan A. Masson et al., 'The use of multiple insulin injection therapy using

"NovoPen" in a routine out-patient setting', *Diabetes Research and Clinical Practice* 7 (1989), pp. 57–60.

48 For example, Jothydev Kesvadev et al., 'Evolution of insulin delivery devices: from syringes, pens, and pumps to DIY artificial pancreas', *Diabetes Therapy* 11 (2020), pp. 1251–69.

49 For the original Scottish research, see John S. Paton et al., 'Convenient pocket insulin syringe', *The Lancet* 317 (1981), pp. 189–90.

50 A short article on the early history of insulin pens, including interview testimony from Reith, is available on Diabetes UK's website; 'The pen is mightier . . .', *Diabetes UK*, <https://www.diabetes.org .uk/about_us/news/diabetes-technology>.

51 This was not, however, a perfect solution. The action profile of prolonged-action insulin available in the 1970s and 1980s continued to have a pronounced peak, meaning perfectly stable basal sugar levels remained difficult to attain. Nonetheless, it was considerably more 'physiological' in pattern than traditional fixed regimens.

52 T.S. Danowski and J.H. Sunder, 'Jet injection of insulin during self-monitoring of blood glucose', *Diabetes Care* 1 (1978), pp. 27–33.

53 Ibid., p. 32.

54 For example, Elizabeth S. McCaughey et al., 'Improved diabetic control in adolescents using the Penject syringe for multiple insulin injections', *Diabetic Medicine* 3 (1986), pp. 234–6.

55 For example, N. Saurbrey et al., 'Comparison of continuous subcutaneous insulin infusion with multiple insulin injections using the NovoPen', *Diabetic Medicine* 5 (1988), pp. 150–3.

56 Masson et al., 'The use of multiple insulin injection therapy', p. 60.

57 K.J. Hardy et al., 'Deterioration in blood glucose control in females with diabetes changed to a basal-bolus regimen using a pen-injector', *Diabetic Medicine* 8 (1991), pp. 69–71.

58 Judith North, 'The pen-injector's impact on self-management', in Clare Bradley, Philip Home, and Margaret Christie (eds), *The Technology of Diabetes Care: Converging Medical and Psychosocial Perspectives* (Chur: Harwood, 1991), p. 84.

59 For example, Bo Feldt-Rasmussen et al., 'Effect of two years of strict metabolic control on progression of incipient nephropathy in insulin-dependent diabetes', *The Lancet* 328 (1986), pp. 1300–4.

60 The DCCT focused specifically on T1DM. However, a few years later the UK Prospective Diabetes Study (UKPDS) demonstrated similar results in T2DM; UKPDS Group, 'Intensive blood-glucose control with sulphonylureas or insulin compared with conventional treatment and risk of complications in patients with Type 2 diabetes (UKPDS 33)', *The Lancet* 352 (1998), pp. 837–53.

61 DCCT Research Group, 'The effect of intensive treatment of diabetes on the development and progression of long-term complications in insulin-dependent diabetes mellitus', *New England Journal of Medicine* 329 (1993), pp. 977–86.

62 Ibid., p. 978.

63 Ibid., pp. 983-4.

64 Stephanie Amiel interviewed by author, 27 February 2019, GB 249 SOHC 64, University of Strathclyde Archives and Special Collections.

65 I. Mühlhauser et al., 'Bicentric evaluation of a teaching and treatment programme for type 1 (insulin-dependent) diabetic patients: improvement of metabolic control and other measures of diabetes care for up to 22 months', *Diabetologia* 25 (1983), p. 470. Assal had been inspired by a period working with Joslin in Boston, and had returned to Switzerland full of zeal for patient education. He discusses the topic in J.P. Assal et al., 'Patient education as the basis for diabetes care in clinical practice and research', *Diabetologia* 28 (1985), pp. 602–13.

66 For more on the experience of diabetes in the context of Soviet-aligned East Germany, see Kathrin A. Hiepko, '"Conditionally healthy and able to work": diabetes prevention, care and research in the German Democratic Republic (GDR), *c.*1949–1990' (unpublished PhD thesis, University of Manchester, 2019).

67 Victor Jörgens, 'A history of patient education for people with diabetes: a very personal view', in Victor Jörgens and Massimo Porta (eds), *Unveiling Diabetes – Historical Milestones in Diabetology* (Basel: Karger, 2020), p. 258.

68 Mühlhauser et al., 'Bicentric evaluation of a teaching and treatment programme', p. 470.

69 Harry Keen, 'Technology and the diabetic patient: an overview', in Clare Bradley, Philip Home, and Margaret Christie (eds), *The Technology of Diabetes Care: Converging Medical and Psychosocial Perspectives* (Chur: Harwood, 1991), p. 10.

70 Mühlhauser et al., 'Bicentric evaluation of a teaching and treatment programme', p. 475.

71 Ibid., p. 471.

72 This was corroborated in a later paper that analysed a much larger sample of 434; Ingrid Mühlhauser et al., 'Incidence and management of severe hypoglycemia in 434 adults with insulin-dependent diabetes mellitus', *Diabetes Care* 8 (1985), pp. 268–73.

73 Mühlhauser et al., 'Bicentric evaluation of a teaching and treatment programme', p. 471.

74 I. Mühlhauser et al., 'Evaluation of an intensified insulin treatment and teaching programme as routine management of Type 1 (insulin-dependent) diabetes: The Bucharest-Düsseldorf Study', *Diabetologia* 30 (1987), pp. 681–90.

75 Ibid., p. 685.

76 Quoted in Jörgens, 'A history of patient education for people with diabetes', p. 260. Edward Tolstoi was aware of Stolte, and cited him as an influence. He appears to have misunderstood the German, however, believing that he cared little about high blood sugar levels; Tolstoi, 'The free diet for diabetic patients', p. 65.

77 Michael Berger, *Bedarfsgerechte Insulin-Therapie bei freier Kost: Der Beitrag von Karl Stolte zur klinischen Diabetologie* (Mainz: Kirchheim, 1999).

78 Amiel, Interview.

79 The final edition was released in 2001, a year before Berger's death; M. Berger and V. Jörgens, *Praxis der Insulintherapie*, 6th edn (Berlin: Springer-Verlag, 2001).

80 Kinga Howorka, *Functional Insulin Treatment*, 1st edn, trans. Kathryn Nelson (Berlin: Springer-Verlag, 1991).

81 Howorka disliked the phrase 'intensive' because it implicitly suggested something 'more onerous and demanding' than traditional management, and suggested the term 'functional' instead; Kinga Howorka, *Functional Insulin Treatment*, 2nd edn, trans. Kathryn Nelson (Berlin: Springer-Verlag, 1996), p. xiii.

82 Ibid., p. 4.

83 Ibid., p. 3.

84 Harry Keen, 'Book review: Functional insulin treatment', *European Journal of Clinical Nutrition* 51 (1997), p. 128.

85 Harry Keen, 'DCCT', *Balance*, August/September 1994, p. 25.

86 Michael Berger, 'Foreword to the first German-language edition', in Kinga Howorka, *Functional Insulin Treatment*, pp. x–xi.

87 Richard K. Bernstein, 'Virtually continuous euglycemia for 5 yr in a labile juvenile-onset diabetic patient under noninvasive closed-loop control', *Diabetes Care* 3 (1980), pp. 140–3.

88 Richard Bernstein, quoted in Penny Singer, 'Diabetic doctor offers a new treatment', *New York Times*, 3 April 1988, p. 12.

89 Jack D. Eastwood, 'Insulin and independence', *BMJ (Clinical Research Edition)* 6562 (1986), p. 1659.

90 Ibid.

91 Anon., 'Obituary: Mr. J.D. Eastwood', *The Craven Herald*, 4 December 1987, reproduced in *Ermysted's Grammar School Chronicles*, Autumn 1987–Summer 1988, pp. 4–5. Courtesy of Ermysted's Grammar School Old Boy's Network.

92 Eastwood self-published *Diabetes Without Tears* – part autobiography, part instruction manual – in 1976; Jack D. Eastwood, *Diabetes Without Tears: A Layman's Account of his Fifty Years on Insulin* (Skipton: Self-Published, 1976).

93 Eastwood, 'Insulin and independence', p. 1660.

94 Fletcher's article was published simultaneously in both the *BMJ* and *Balance*, with some adaptations to reflect their different readerships; Charles Fletcher, 'An active diabetic life', *Balance*, December 1980, pp. 1, 15; Charles Fletcher, 'One way of coping with diabetes', *BMJ* 6222 (1980), pp. 1115–16. For one example of Fletcher's critique of the traditional paternalistic relationship, see Charles M. Fletcher, 'Communication between doctors and patients', *Proceedings of the Royal Society of Medicine* 61 (1968), pp. 567–8.

95 'Ms E.J.', 'Concern', *Balance*, April 1981, p. 2; 'J.B.P', 'Horror', *Balance*, February 1981.

Chapter 4: Subjectivity, Paternalism, Neoliberalism, 1993–2002

1 DAFNE Study Group, 'Training in flexible, intensive insulin management to enable dietary freedom in people with Type 1 diabetes: dose adjustment for normal eating (DAFNE) randomised controlled trial', *BMJ* 7367 (2002), 10.1136/bmj.325.7367.746

2 DAFNE is, however, the source of some controversy here. The Beta Cell Education Resources for Training in Insulin and Eating

(BERTIE) programme, with which it has something of a long-running rivalry, has some claim to precedence. BERTIE originated in the late 1990s, was grounded in similar principles, and was derived from the same influences as DAFNE. Its organizers, however, did not publish their initial results in the peer-reviewed literature, so it – perhaps unfairly – received comparatively little attention; L. Grant et al., 'Type 1 diabetes structured education: what are the core self-management behaviours?', *Diabetic Medicine* 30 (2013), p. 727.

3 China, as an example, has recently begun to experiment with the principle; Yuting Xie et al., 'Establishment of a type 1 diabetes structured education programme suitable for Chinese patients: type 1 diabetes education in lifestyle and self adjustment (TELSA)', *BMC Endocrine Disorders* 20 (2020), https://doi.org/10.1186/s12902-020-0514-9.

4 Follow-up studies have, however, highlighted that 'refresher' courses might be useful. While they remain lower than the baseline average for people with T1DM, participants' HbA1c values do appear to gradually rise several years after attending the initial DTTP; J. Speight et al., 'Long-term biomedical and psychosocial outcomes following DAFNE (Dose Adjustment for Normal Eating) structured education to promote intensive insulin therapy in adults with suboptimally controlled Type 1 diabetes', *Diabetes Research and Clinical Practice* 89 (2010), pp. 22–9.

5 Amiel, Interview; Simon Heller interviewed by author, 3 July 2019, GB 249 SOHC 64, University of Strathclyde Archives and Special Collections.

6 Dorothy Holland and Kevin Leander, 'Ethnographic studies of positioning and subjectivity: an introduction', *Ethos* 32 (2004), p. 129.

7 Emphasis in original; T.M. Luhrmann, 'Subjectivity', *Anthropological Theory* 6 (2006), p. 346.

8 In addition to physical disorders, this theme holds a significant amount of relevance for mental health and neurodevelopmental conditions, the precise boundaries of which are the subject of ongoing (and tricky to resolve) debate. For one example of a work engaging with these topics – in this case focusing on autism – see Mitzi Waltz, *Autism: A Social and Medical History* (Basingstoke: Palgrave Macmillan, 2013).

9 Jay Katz, *The Silent World of Doctor and Patient* (Baltimore, MD: Johns Hopkins University Press, 1984), pp. 90–1.

10 Talcott Parsons, *The Social System* (New York, NY: Free Press of Glencoe, 1951), p. 285.

11 Ibid., p. 441.

12 Winn, *Diabetes Stories*.

13 Clifton, Interview.

14 Vic Marriott interviewed by author, 9 May 2017, GB 249 SOHC 64, University of Strathclyde Archives and Special Collections.

15 Frank Kaye and Sylvia Kaye interviewed by author, 22 June 2017, GB 249 SOHC 64, University of Strathclyde Archives and Special Collections.

16 Marriott, Interview.

17 Cowan, Interview.

18 This is not, of course, a static process. What constitutes 'typical' is wholly culturally constructed; Paul Chatterton, 'University students and city centres – the formation of exclusive geographies: the case of Bristol, UK', *Geoforum* 30 (1999), pp. 117–33.

19 Balfe takes the term 'body project' from Chris Shilling; Chris Shilling, *The Body and Social Theory*, 2nd edn (London: SAGE Publications, 2003).

20 Myles Balfe, 'The body projects of university students with Type 1 diabetes', *Qualitative Health Research* 19 (2009), pp. 128–39; Myles Balfe, 'Healthcare routines of university students with Type 1 diabetes', *Journal of Advanced Nursing* 65 (2009), pp. 2367–75.

21 Shelley Budgeon, 'Identity as an embodied event', *Body and Society* 9 (2003), pp. 35–55.

22 Josiah D. Rich et al., 'Insulin use by bodybuilders', *Journal of the American Medical Association* 279 (1998), p. 1613; P.J. Evans and R.M. Lynch, 'Insulin as a drug of abuse in body building', *British Journal of Sports Medicine* 37 (2003), 356–7; Lindsey J. Anderson et al., 'Use of growth hormone, IGF-I, and insulin for anabolic purpose: pharmacological basis, methods of detection, and adverse effects', *Molecular and Cellular Endocrinology* 464 (2018), pp. 65–74.

23 'Generation iron fitness and bodybuilding network', 4 June 2021, 'Milos Sarcev Full Interview | Steroids, Insulin, & PEDs in Bodybuilding', *YouTube*, <https://youtu.be/CDQHfxYSqXQ>.

24 For one example of a doctor directly addressing the bodybuilding community and encouraging them to reject insulin use, see 'Anabolic Doc',

21 May 2019, 'Insulin for bodybuilding? – Doctor's analysis of side effects and properties', *YouTube*, <https://youtu.be/Wwva9Zj1x8I>.

25 For example, Maria Vultaggio, 'Dallas McCarver's 911 call reveals insulin possibly connected to cause of death', *International Business Times*, 24 August 2017, <https://www.ibtimes.com/dallas-mccarvers-911-call-reveals-insulin-possibly-connected-cause-death-2582322>.

26 'Milos Sarcev Full Interview | Steroids, Insulin, & PEDs in Bodybuilding', *YouTube*.

27 For example, 'Leo and Longevity', 'Debating Insulin Use || Vigorous Steve & Bostin Loyd', 16 June 2021, *YouTube*, <https://youtu.be/5O4uqrMCA78>.

28 For example, Taeian Clark, 'The real 101 guide to insulin and bodybuilding – doses, hyperplasia, timing', *Taeian.com Blog*, 10 July 2018, <https://www.taeian.com/the-real-101-guide-to-insulin-and-bodybuilding-doses-hyperplasia-timing/>.

29 Evans and Lynch, 'Insulin as a drug of abuse in body building', p. 356.

30 Collins and Pinch draw on an earlier study by sociologist Lee Monaghan; Harry Collins and Trevor Pinch, *Dr. Golem: How To Think About Medicine* (Chicago, IL: Chicago University Press, 2005), pp. 125–7; Lee F. Monaghan, *Bodybuilding, Drugs, and Risk* (London: Routledge, 2000).

31 Collins and Pinch, *Dr. Golem*, p. 127.

32 Unsurprisingly, psychiatry was the focus of many early works of this kind. Michel Foucault, for example, was one of the first to address the subject, and he went on publish work discussing the power structures of the medical profession more generally; Michel Foucault, *Madness and Civilization: A History of Insanity in the Age of Reason [1961]*, trans. Richard Howard (London: Tavistock, 1967); Michel Foucault, *The Birth of the Clinic: An Archaeology of Medical Perception [1963]*, trans. A.M. Sheridan (London: Tavistock, 1973). Foucault can, however, be quite esoteric. In 1969, Thomas Szasz wrote in much more straightforward terms that the 'paternalistic psychiatrist . . . assumes "responsibility" for [the person under their care], defines him as a "patient" against his will, and subjects him to "treatment" deemed best for him, with or without his consent'; Thomas Szasz, *Ideology and Insanity: Essays on the Psychiatric Dehumanization of Man* (Garden City, NY: Doubleday Anchor, 1969), p. 86.

33 For two of his most influential works, see Arthur Kleinman, *Patients and Healers in the Context of Culture: An Exploration of the Borderland between Anthropology, Medicine, and Psychiatry* (Berkeley, CA: University of California Press, 1980); Arthur Kleinman, *The Illness Narratives: Suffering, Healing, and the Human Condition* (New York: Basic Books, 1988).

34 Simon R. Heller, 'Robert Tattersall, a diabetes physician ahead of his time', *Diabetes Care* 42 (2010), https://doi.org/10.2337/dci18-0063.

35 Michal Anděl and Robert Tattersall, 'Authoritarianism in diabetology', *Diabetic Medicine* 6 (1989), p. 471.

36 Robert Tattersall, 'Reply from Tattersall', *Diabetic Medicine* 6 (1989), pp. 830–1. For more on the concept of 'noncompliance' in medicine, see Jeremy A. Greene, 'Therapeutic infidelities: 'noncompliance' enters the medical literature, 1955–1975', *Social History of Medicine* 17 (2004), pp. 327–43.

37 Margaret Howie interviewed by author, 27 July 2017, GB 249 SOHC 64, University of Strathclyde Archives and Special Collections.

38 'N.D.', 'Bending the rules', *Balance*, August 1977, p. 2.

39 Alex Mold, 'Patients' rights and the National Health Service in Britain, 1960s–1980s', *American Journal of Public Health* 102 (2012), p. 2032.

40 David Kelleher, 'Non-compliance and the toleration of symptoms', *Balance*, August/September 1987, p. 27.

41 Alexandra Weston, 'Non-compliance', *Balance*, October/November 1987, p. 7.

42 Kelleher, 'Non-compliance and the toleration of symptoms', p. 27. The following year, Kelleher published a more substantial book on the topic; David Kelleher, *Diabetes* (London: Routledge, 1988).

43 Rowan Hillson, *Diabetes: A Beyond Basics Guide*, 2nd edn (London: Optima, 1992), p. 25.

44 For another example of a (slightly later) article expressing similar opinions, see Peter G.F. Swift, 'Flexible carbohydrate', *Diabetic Medicine* 14 (1997), pp. 187–8.

45 Amiel, Interview.

46 Hillson, *Diabetes*, p. 64.

47 T. Deckert et al., 'Natural history of diabetic complications: early detection and progression', *Diabetic Medicine* 8 *(Supplement 2)* (1991), pp. 33–7.

48 'Diabetes care and research in Europe: the Saint Vincent Declaration', *Diabetic Medicine* 7 (1990), p. 360.

49 Harry Keen, 'The St Vincent Declaration – origins and perspectives', *Practical Diabetes International* 17 (2000), pp. 211–12.

50 Audit Commission for Local Authorities and the National Health Service in England and Wales, *Testing Times: A Review of Diabetes Services in England and Wales* 2000).

51 E.G. Starostina et al., 'Effectiveness and cost-benefit analysis of intensive treatment and teaching programmes for Type 1 (insulin-dependent) diabetes mellitus in Moscow – blood glucose versus urine glucose self-monitoring', *Diabetologia* 37 (1994), pp. 170–6.

52 Michael Berger and Ingrid Mühlhauser, 'Implementation of intensified insulin therapy: a European perspective', *Diabetic Medicine* 12 (1995), p. 206.

53 'Breakthrough offers freedom for people with diabetes', *Diabetes Today* 5 (2002), p. 119; David Jack, 'DAFNE – controlling diabetes the German way', *The Lancet* 357 (2001), p. 1185.

54 Amiel, Interview.

55 Heller, Interview.

56 While DAFNE was specifically designed for T1DM, similar programmes were soon implemented that concentrated on T2DM. For example, T.C. Skinner et al., 'Diabetes education and self-management for ongoing and newly diagnosed (DESMOND): process modelling of pilot study', *Patient Education and Counseling* 64 (2006), pp. 369–77.

57 For example, Anne Kinch et al., 'Local implementation of a carbohydrate counting system', *Journal of Diabetes Nursing* 8 (2004), pp. 28–30.

58 For example, Paul Dromgoole, 'Supporting people with diabetes to engage with insulin therapy', *Journal of Diabetes Nursing* 18 (2014), pp. 378–84.

59 For example, in Aled Davies, Ben Jackson, and Florence Sutcliffe-Braithwaite (eds), *The Neoliberal Age? Britain since the 1970s* (London: UCL Press, 2021).

60 Stephanie Lee Mudge, 'What is neo-liberalism?', *Socio-Economic Review* 6 (2008), pp. 704–5.

61 Mark Fisher, *Capitalist Realism: Is There No Alternative?* (Winchester: Zero Books, 2009), p. 8.

62 David Harvey, *A Brief History of Neoliberalism* (Oxford: Oxford University Press, 2005), pp. 2–3.

63 Alex Mold, 'Making the patient-consumer in Margaret Thatcher's Britain', *The Historical Journal* 54 (2011), pp. 509–28.

64 Department of Health, *Patient's Charter* (1991).

65 Mark Wickham-Jones, 'Neoliberalism and the Labour Party', in Aled Davies, Ben Jackson, and Florence Sutcliffe-Braithwaite (eds), *The Neoliberal Age? Britain since the 1970s* (London: UCL Press, 2021), p. 227.

66 New Labour's health policy has been the subject of much literature in and of itself. For one excellent history, see Rudolf Klein, *The New Politics of the NHS: from Creation to Reinvention*, 7th edn (London: CRC Press, 2013). Unsurprisingly, the subject has also produced considerable amounts of more pointedly polemical – though no less necessary – work, such as John Lister, *The NHS after 60: For Patients or Profits?* (London: Middlesex University Press, 2008).

67 Peter Moore, 'Type 2 diabetes is a major drain on resources', *BMJ* 7237 (2000), p. 732.

68 Department of Health, *The Expert Patient: A New Approach to Chronic Disease Management for the 21st Century* (2001), p. 5.

69 Ibid., p. 26.

70 Sara M. Glasgow, 'The politics of self-craft: expert patients and the public health management of chronic disease', *SAGE Open* 2 (2012), https://doi.org/10.1177%2F2158244012452575.

71 Joanne Shaw and Mary Baker, '"Expert patient" – dream or nightmare?', *BMJ* 7442 (2004), p. 723.

72 Ibid., p. 724.

73 N.J. Fox et al., 'The "expert patient": empowerment or medical dominance? The case of weight loss, pharmaceutical drugs and the Internet', *Social Science and Medicine* 60 (2005), p. 1300.

74 Patricia Wilson, 'A policy analysis of the expert patient in the United Kingdom: self-care as an expression of pastoral power', *Health and Social Care in the Community* 9 (2001), p. 134.

75 Amiel, Interview.

76 This point was highlighted in Howard A. Wolpert and Barbara J. Anderson, 'Management of diabetes: are doctors framing the benefits from the wrong perspective?', *BMJ* 7319 (2001), pp. 994–6.

77 Heller, Interview.

78 For example, Anon., 'The DAFNE course absolutely blew me away', *Diabetes UK*, <https://www.diabetes.org.uk/your-stories/type-1/the -dafne-course-absolutely-blew-me-away>; Various, 'DAFNE – is it worth it!?', *Diabetes.co.uk*, 4 October 2015 <https://www.diabetes.co .uk/forum/threads/dafne-is-it-worth-it.108920/>.

79 For example, 'Fearless36', 'DAFNE – For me – didn't work', *Diabe tes.co.uk*, 21 June 2020, <https://www.diabetes.co.uk/forum/threads /dafne-for-me-didnt-work.175165/>.

80 'Smidge', 'DAFNE experience – the good and the bad!', *Diabetes.co.uk*, 26 January 2014, <https://www.diabetes.co.uk/forum/threads/dafne-experience-the-good-and-the-bad.52066/>.

81 Heller, Interview.

82 Berger, 'Foreword', in Howorka, *Functional Insulin Treatment*, p. x.

83 Helen Hopkinson, 'Why should you commission structured patient education for adults with type 1 diabetes?', *British Journal of Diabetes* 16 (2016), pp. 101–2; NICE, *Type 1 diabetes in adults: diagnosis and management [NG17]* (2015).

84 Glasgow, 'The politics of self-craft'.

85 Initially, for example, CSII was reserved only for those who could not 'control their diabetes by other methods'; John Davis and Valerie Wilson, 'NICE: the way forward with insulin pumps', *Diabetes and Primary Care* 6 (2004), p. 72.

86 North Bristol NHS Trust Insulin Pump Service, 'Agreement of responsibilities for users of NHS funded insulin pumps', *Association of British Clinical Diabetologists* <https://abcd.care/dtn/appendices-dtn-ser vice-best-practice-guide>.

87 Rabab et al., 'A three-way accuracy comparison of the Dexcom G5, Abbott Freestyle Libre Pro, and Senseonics Eversence continuous glucose monitoring devices in a home-use study of subjects with Type 1 diabetes', *Diabetes Technology and Therapeutics* 22 (2022), pp. 846–52.

88 L. Leelarathna and E.G. Wilmot, 'Flash forward: a review of flash glucose monitoring', *Diabetic Medicine* 35 (2018), pp. 462–82.

89 All emphasis in original; @Tims_Pants ('Tim Street'), 'If you're in Birmingham, additional criteria have been added to #Libre trial peri ods that make life more work . . .', *Twitter*, 28 April 2021, https:// twitter.com/Tims_Pants/status/1387319267641659393.

90 For example, 'Mattrix', 'DAFNE – spoilers please!', *Diabetes.co.uk*, 22 April 2022, <https://www.diabetes.co.uk/forum/threads/dafne -spoilers-please.173783/>.

91 'ElyDave', 'DAFNE experience – the good and the bad!', *Diabetes .co.uk*, 27 January 2014, <https://www.diabetes.co.uk/forum/threa ds/dafne-experience-the-good-and-the-bad.52066/>.

92 Quoted in Dympna Casey et al., 'A mixed methods study exploring the factors and behaviours that affect glycemic control following a structured education program: the Irish DAFNE Study', *Journal of Mixed Methods Research* 10 (2014), https://doi.org/10.1177/155868 9814547579.

93 For example, 'Bluemarie Josephine', 'DAFNE Confusion', *Diabetes .co.uk*, 5 November 2015, <https://www.diabetes.co.uk/forum/thre ads/dafne-confusion.86977/>.

94 For example, Allan S. Berger, 'Arrogance among physicians', *Academic Medicine* 77 (2002), pp. 145–7.

95 Amélie Oksenberg Rorty, 'The vanishing subject: the many faces of subjectivity', in João Biehl, Byron Good, and Arthur Kleinman (eds), *Subjectivity* (Berkeley, CA: University of California Press, 2007), p. 48.

Chapter 5: The Insulin Crisis, 2002–Present

1 Tiffany Stanley, 'Life, death and insulin', *Washington Post*, 7 January 2019, <https://www.washingtonpost.com/news/magazine/wp/2019 /01/07/feature/insulin-is-a-lifesaving-drug-but-it-has-become-in tolerably-expensive-and-the-consequences-can-be-tragic/>.

2 Medicare and Medicaid provide financial assistance to the elderly and permanently disabled, and the extremely poor respectively.

3 The ACA has been roundly criticized because it fell far short of ensuring universal access to healthcare. Nonetheless, it represents one of the most radical overhauls of the American healthcare system to date. For one critical analysis, see Laxmaiah Manchikanti et al., 'A critical analysis of Obamacare: affordable care or insurance for many and coverage for few?', *Pain Physician* 20 (2017), pp. 111–38.

4 The 'out of pocket maximum' of the policy Smith was looking at is unknown. Sometimes this is set at the same rate as the deductible, but in many cases it is significantly higher. Both are generally much higher

in policies with lower monthly premiums – often the only policies the less wealthy can afford. As ever, it is expensive to be poor.

5 The article in question does, I feel, somewhat miss the mark, however, in its suggestion that this could be accomplished via further competition in the manufacturing market; Dzintars Gotham et al., 'Production costs and potential prices for biosimilars of human insulin and insulin analogues', *BMJ Global Health* 3 (2019), http://dx.doi.org/10.1136/bmjgh-2018-000850.

6 Nicole Hood, 'I lost Allen because he was forced to ration insulin', *T1International*, 6 July 2019, <https://www.t1international.com/blog/2019/07/06/rationing-insulin-caused-andys-death/>.

7 Jazmine Baldwin, 'My sister Jada should never have died', *T1International*, 18 July 2019, <https://www.t1international.com/blog/2019/07/18/my-sister-jada-should-never-have-died/>.

8 Janelle Lutgen, 'Remembering Jesse in the fight for #insulin4all', *T1International*, 1 July 2019, <https://www.t1international.com/blog/2019/07/01/remembering-jesse-fight-insulin4all/>.

9 Irl B. Hirsch, 'Insulin in America: a right or a privilege?', *Diabetes Spectrum* 29 (2016), pp. 130–2; Samantha Willner et al., '"Life or death": experiences of insulin insecurity among adults with Type 1 diabetes in the United States', *SSM – Population Health* 11 (2020), https://dx.doi.org/10.1016%2Fj.ssmph.2020.100624.

10 Arleen Marcia Tuchman, *Diabetes: A History of Race and Disease* (New Haven, CT: Yale University Press, 2020).

11 For example, Albert A. Epstein, 'Diabetes among Jews – its cause and prevention', *Modern Medicine* 1 (1919), p. 20; Arthur R. Elliott, 'Diabetes mellitus: the limitations of its dietetic treatment', *Journal of the American Medical Association* 42 (1904), pp. 19–23.

12 For example, Emil Kleen, *On Diabetes Mellitus and Glycosuria* (Philadelphia, PA: P. Blakiston's, Son & Co, 1900).

13 For example, Robert Saundby, 'An address on the modern treatment of diabetes mellitus', *The Lancet* 155 (1900), pp. 1420–6.

14 George M. Beard, *American Nervousness: Its Causes and Consequences* (New York, NY: G.P. Putnam's Sons, 1881), p. 59.

15 Ibid., p. 130.

16 Ibid., pp. 130–1.

17 Ibid., p. 131. Interestingly, actual Indians – people from the Indian

subcontinent – were not generally thought of in this way. In fact, they were often considered particularly susceptible. For more on this, see David Arnold, 'Diabetes in the tropics: race, place and class in India, 1880–1965', *Social History of Medicine* 22 (2009), pp. 245–61.

18 Tellingly, those working-class people who did develop diabetes rarely benefited from such an interpretation; Shane O'Donnell, 'Changing social and scientific discourse on Type 2 diabetes between 1800 and 1950: a socio-historical analysis', *Sociology of Health and Illness* 37 (2015), pp. 1109–10.

19 Charles W. Purdy, *Diabetes: Its Causes, Symptoms, and Treatment* (Philadelphia, PA: F.A. Davis, 1890), p. 18.

20 This stereotype was occasionally embraced by Jewish people themselves – highlighted as evidence of the generational trauma produced by centuries of oppression and sometimes outright brutality. For example, Maurice Fishberg, 'Health and sanitation of the immigrant Jewish population of New York', *The Menorah: A Monthly Magazine for the Jewish Home* 33 (1902), p. 170.

21 Reginald H. Fitz and Elliott P. Joslin, 'Diabetes mellitus at the Massachusetts General Hospital from 1824 to 1898: a study of the medical records', *Journal of the American Medical Association* 31 (1898), p. 171.

22 Haven Emerson, 'Sweetness is death', *The Survey* 53 (1924), p. 25.

23 For example, J.H. Parks and E. Waskow, 'Diabetes among the Pima Indians of Arizona', *Arizona Medicine: Journal of the Arizona State Medical Association* 18 (1961), pp. 99–107.

24 The concept of a 'thrifty gene' was first described by James V. Neel in 1962. He did not initially, however, draw any connection to Indigenous communities; James V. Neel, 'Diabetes mellitus: a "thrifty" genotype rendered detrimental by "progress"?', *American Journal of Human Genetics* 14 (1962), pp. 353–62.

25 For example, P.H. Wise et al., 'Diabetes and associated variables in the South Australian Aboriginal', *Australian and New Zealand Journal of Medicine* 6 (1976), pp. 191–6; Paul Zimmet, 'Epidemiology of diabetes and its macrovascular manifestations in Pacific populations: the medical effects of social progress', *Diabetes Care* 2 (1979), pp. 144–53; William C. Knowler et al., 'Diabetes incidence in Pima Indians: contributions of obesity and parental diabetes', *American Journal of Epidemiology* 113 (1981), pp. 144–56.

26 US Department of Health and Human Services, *Report of the Secretary's Task Force on Black and Minority Health, Volume VII: Chemical Dependency and Diabetes* (1985), p. 193.

27 Elliott Joslin, 'The universality of diabetes: a survey of diabetic mortality in Arizona. The Billings Lecture', *Journal of the American Medical Association* 115 (1940), pp. 2033–8.

28 Louis Dublin, 'The health of the Negro', *Annals of the American Academy of Political and Social Science* 140 (1928), pp. 77–85.

29 'Statistics about diabetes', *ADA*, 4 February 2022, <https://www.diabetes.org/about-us/statistics/about-diabetes>.

30 'UK's poorest twice as likely to have diabetes and its complications', *Diabetes UK*, 27 August 2009, <https://www.diabetes.org.uk/about_us/news_landing_page/uks-poorest-twice-as-likely-to-have-diabetes-and-its-complications>; 'Diabetes statistics', *Diabetes UK* <https://www.diabetes.org.uk/professionals/position-statements-reports/statistics>.

31 Tuchman, *Diabetes*, pp. 188–9.

32 Stephen J. Dubner, "The most bountiful food in human history?", *Freakonomics*, 21 March 2013, <https://freakonomics.com/2013/03/the-most-bountiful-food-in-human-history/>.

33 Lora Arduser, 'What's in a name? The diabetes civil war', in Bianca C. Frazer and Heather R. Walker (eds), *(Un)doing Diabetes: Representation, Disability, Culture* (Chur: Palgrave, 2021), pp. 43–62.

34 Jessica L. Browne et al., '"I'm not a druggie, I'm just a diabetic": a qualitative study of stigma from the perspective of adults with type 1 diabetes', *BMJ Open* 4 (2014), https://doi.org/10.1136/bmjopen-2014-005625.

35 Ludwig has since deleted this Tweet following significant backlash, but the article to which it linked can be found at David Ludwig, 'Back to the future for diabetes', *Better Humans*, 20 October 2021, <https://betterhumans.pub/back-to-the-future-for-diabetes-a7fdf713856>.

36 It seems particularly tone-deaf that the article was published as part of the journal's '100th Anniversary of Insulin's Discovery' review series; Belinda S. Lennerz et al., 'Carbohydrate restriction for diabetes: rediscovering centuries-old wisdom', *Journal of Clinical Investigation* 131 (2021), https://doi.org/10.1172/jci142246.

37 @KalChinyereMD ('Dr. Kal Chinyere'), 'But, if we cut their insulin

requirement in half, don't we cut their costs in half?', *Twitter*, 21 October 2021, https://twitter.com/KalChinyereMD/status/145123 0729334050818?s=20&t=qLXo83TbshMAfdgRFpaxpw.

38 Studies continue to show that diabetes shortens lives considerably; for example, Shona J. Livingstone et al., 'Estimated life expectancy in a Scottish cohort with Type 1 diabetes, 2008–2010', *Journal of the American Medical Association* 313 (2015), pp. 37–44. However, there is some space to be hopeful here. The pace of change in the way insulin therapy has been conducted since the beginning of the twenty-first century limits the ability of such research to accurately assess the long-term influence of more recent developments on longevity.

39 There are some smaller insulin manufacturers across the globe, but they are collectively dwarfed by the 'Big Three'; Veronika J. Wirtz, *Insulin Market Profile* (Amsterdam: Health Action International, 2016), p. 20.

40 Mike Hoskins, 'Insulin availability for those who need it most (remembering Shane Patrick Boyle)', *Healthline*, 3 April 2017, <https://www.healthline.com/diabetesmine/insulin-access-deaths>.

41 There is also some scepticism amongst people with diabetes about the $7,500 figure. For most co-pay cards, this is a lifetime limit, not an annual one. While Eli Lilly insists that this is not the case in this instance, some have reported difficulties when trying to renew for a second year; for example @FennerMichelle ('Michelle Fenner'), 'I was on Otezla with a 9000 lifetime deductible', 11 April 2020, *Twitter*, https://twitter.com/FennerMichelle/status/1249034293382389760.

42 Rachel Balick, 'Copay cards save patients money, but come at a cost', *Pharmacy Today* 22 (2016), https://doi.org/10.1016/j.ptdy.2016.11.023.

43 Katherine Kraschel and Gregory Curfman, 'Patient assistance programs and anti-kickback laws', *Journal of the American Medical Association* 322 (2019), pp. 405–6; David Lazarus, 'Column: Drug companies are growing less generous in helping patients pay for meds', *Los Angeles Times*, 15 August 2017, <https://www.latimes.com/business/lazarus/la-fi-lazarus-prescription-drug-assistance-20170815-story.html>.

44 For a selection of (bad) arguments, including this one, from manufacturers, see 'Insulin makers respond to outrage over skyrocketing prices', *Healthline*, 1 June 2018, <https://www.healthline.com/diabetesmine/high-insulin-price-response>.

45 Michael F. Cannon, the director of health policy studies for the libertarian Cato Institute, for example, is very keen on this argument; Michael F. Cannon, 'The last thing insulin markets need is more government', 10 August 2022, *Cato Institute*, <https://www.cato.org/commentary/last-thing-insulin-markets-need-more-government>. Cannon has form for some quite absurd positions on this topic, however. In one recent social media post, he suggested that making insulin free of charge to consumers would create shortages, because, implicitly, they would use more of it and exhaust the supply. Having enjoyed NHS-funded prescriptions for my entire life in the UK, I would venture to suggest that he may be mistaken. One can only imagine the panic he might feel should someone point out to him that oxygen is also free, and people are able to breathe as much of it as they like; @mfcannon ('Michael F. Cannon and 14398 others'), 'Nationalising insulin production would cause its cost to rise by suppressing innovation', 10 August 2022, *Twitter*, https://twitter.com/mfcannon/status/1557435895204159488.

46 The development of 'human' insulin is described at length in Stephen S. Hall, *Invisible Frontiers: The Race to Synthesize a Human Gene* (London: Sidgwick & Jackson, 1988). While there may be a good horror movie in the idea, this is not, as the name might suggest, made from humans. It is in fact completely synthetic, and produced under laboratory conditions.

47 Jeremy A. Greene and Kevin R. Riggs, 'Why is there no generic insulin? Historical origins of a modern problem', *New England Journal of Medicine* 372 (2015), pp. 1171–5.

48 Older animal-derived insulin is still sold for veterinary use, and some desperate people with diabetes have taken to buying these cheaper products to treat themselves; Alan MacLeod, 'Injecting yourself with dog insulin? Just a normal day in America', *The Guardian*, 1 August 2019, <https://www.theguardian.com/commentisfree/2019/aug/01/us-healthcare-insulin-diabetes-jordan-williams>.

49 Stephen C.L. Gough, 'A review of human and analogue insulin trials', *Diabetes Research and Clinical Practice* 77 (2007), doi:10.1016/j.diabres.2006.10.015

50 'Novolog', *RxList*, 17 March 2022, <https://www.rxlist.com/novolog-drug/patient-images-side-effects.htm>.

51 'Lantus', *RxList*, 23 March 2021, <https://www.rxlist.com/lantus-drug.htm>.

52 K. Howorka et al., 'Dealing with ceiling baseline treatment satisfaction level in patients with diabetes under flexible, functional insulin treatment: assessment of improvements in treatment satisfaction with a new insulin analogue', *Quality of Life Research* 9 (2000), pp. 915–30.

53 Israel Hartman, 'Insulin analogs: impact on treatment success, satisfaction, quality of life, and adherence', *Clinical Medicine and Research* 6 (2008), pp. 54–67.

54 This echoed an earlier episode in the 1960s, when the East German government received criticism for clamping down on insulin imports from the West. While the GDR did make some of its own, it lacked the extended-action varieties that many had come to use; Kathrin A. Hiepko, 'Where is the Hoechst insulin? The role of diabetics and their doctors as consumers during the German Democratic Republic's autarkic policy of "making free from disturbance", 1961–1966', *Social History of Medicine* 33 (2020), pp. 924–45.

55 F. Holleman and E.A. Gale, 'Nice insulins, pity about the evidence', *Diabetologia* 50 (2007), pp. 1783–90.

56 S.V. Rajkumar, 'The high cost of insulin in the United States: an urgent call to action', *Mayo Clinic Proceedings* 95 (2020), pp. 22–8.

57 'How much should I expect to pay for Humalog U-100®?', *Eli Lilly*, <https://www.lillypricinginfo.com/humalog>.

58 @Kidfears99 ('Laura Marston'), 'This isn't accurate', *Twitter*, 19 January 2021, https://twitter.com/Kidfears99/status/13514866499 85867777.

59 N.J. Goldstein et al., 'Frequency of sale and reasons for purchase of over-the-counter insulin in the United States', *Journal of the American Medical Association – Internal Medicine* 179 (2019), pp. 722–3.

60 For example, Elizabeth Snouffer, 'ReliOn insulin dangerous for Type 1 diabetes', *Diabetes Voice*, 7 August 2019, <https://diabetesvoice.org/en/diabetes-views/relion-insulin-dangerous-for-type-1-diabetes/>.

61 For example, Ginger Vieira, 'Everything you need to know about Walmart insulin', *DiabetesStrong*, 24 January 2020, <https://diabetesstrong.com/walmart-insulin>.

62 For one news story about someone dying after switching to 'Walmart insulin', see Antonio Olivo, 'He lost his insurance and turned to a cheaper form of insulin. It was a fatal decision', *Washington Post*, 3 August 2019, https://www.washingtonpost.com/local/he-lost-his-insu

rance-and-turned-to-cheaper-form-of-insulin-it-was-a-fatal-decision/
2019/08/02/106ee79a-b24d-11e9-8f6c-7828e68cb15f_story.html>.

63 'Our history', *The diaTribe Foundation* <https://diatribe.org/foun
dation/about-us/our-history>.

64 Frida Velcani, 'Uninsured and need insulin?', *The diaTribe Foundation*,
20 May 2020, *<https://diatribe.org/uninsured-and-need-insulin>*.

65 While an impressive scientific breakthrough, several studies have
indicated that the difference between 'human' insulin and older
varieties is negligible in practice; B. Richter and G. Neises, '"Human"
insulin versus animal insulin in people with diabetes mellitus',
Cochrane Database of Systematic Reviews (2005), doi: 10.1002/14651858.
CD003816.pub2.

66 Tony Huzzey, *My Life with Diabetes: 61 Years of Carb Counting*
(Rothersthorpe: Paragon, 2011), pp. 191–2.

67 Many studies were published on this subject. One of the first was A.
Teuscher and W.G. Berger, 'Hypoglycaemia unawareness in diabetics
transferred from beef/porcine insulin to human insulin', *The Lancet*
330 (1987), pp. 382–5. For a later example, see M. Egger et al., 'Risk
of severe hypoglycaemia in insulin treated diabetic patients transferred
to human insulin: a case control study', *BMJ* 6803 (1991), pp. 617–21.

68 'Human insulin', *BMJ* 6706 (1989), pp. 991–3.

69 'Walmart revolutionizes insulin access & affordability for patients
with diabetes with the launch of the first and only private brand
analog insulin', *Walmart*, 29 June 2021, <https://corporate.walmart
.com/newsroom/2021/06/29/walmart-revolutionizes-insulin-access-
affordability-for-patients-with-diabetes-with-the-launch-of-the-first-
and-only-private-brand-analog-insulin>.

70 Miriam E. Tucker, 'Walmart's new insulin still too expensive,
advocates say', *WebMD*, 2021, <https://www.webmd.com/diabetes
/news/20210701/walmarts-new-insulin-still-too-expensive-advocates-
say>.

71 Federal Bureau of Prisons, *Management of Diabetes* (2017), https://
www.bop.gov/resources/pdfs/diabetes_guidance_march_2017.pdf>.

72 'Diabetes behind bars: challenging inadequate care in prisons', *Lancet
Diabetes and Endocrinology* 6 (2018), p. 347.

73 Andrew Emett, 'Former jail administrator sentenced to prison for
depriving insulin to diabetic inmate', *Nation of Change*, 11 August 2017,

<https://www.nationofchange.org/2017/08/11/former-jail-administra
tor-sentenced-prison-depriving-insulin-diabetic-inmate/>.

74 Allen S. Keller et al., 'Diabetic ketoacidosis in prisoners without access
to insulin', *Journal of the American Medical Association* 269 (1993),
pp. 619–21. This is far from a uniquely American problem. For example, 'Diabetic "denied medication" in police custody', *The Scotsman*,
16 September 2015, <https://www.scotsman.com/regions/diabetic-
denied-medication-police-custody-1495286>.

75 For one example of someone highlighting this on social media, see @
KVHuntley ('Prof. Katherine V. Huntley'), 'Since I am overdue for an
appointment w/ my endocrinologist by 1 month, they won't write me
a new Rx for my insulin unless I make an appt', *Twitter*, 25 May 2020,
https://twitter.com/KVHuntley/status/1242897154118578176.

76 Ann Socolofsky, 'Is it normal for a primary care doctor to refuse a refill
of a diabetic medication (non-narcotic) because he wants to see his
patient first?', *Quora*, *c*.2019, <https://www.quora.com/Is-it-normal-f
or-a-primary-care-doctor-to-refuse-a-refill-of-a-diabetic-medication-
non-narcotic-because-he-wants-to-see-his-patient-first>.

77 'NHS restricts essential diabetes equipment', *Diabetes UK*, 4 June
2017, <https://www.diabetes.org.uk/in_your_area/wales/news/nhs-res
tricts-essential-diabetes-equipment-in-wales>.

78 Nancy J.V. Bohannon, 'Insulin delivery using pen devices', *Postgraduate
Medicine* 106 (1999), p. 58.

79 'Forced nonmedical switching', *ADA*, <https://www.diabetes.org
/healthy-living/medication-treatments/insurance-switching-diabetes-
medication>.

80 For example, Tyler Choi, 'American caravan arrives in Canadian
"birthplace of insulin" for cheaper medicine', *Reuters*, 29 June 2019,
<https://www.reuters.com/article/us-canada-health-insulin-idUSK
CN1TU0T4>.

81 'Age-adjusted comparative prevalence of diabetes', *IDF Diabetes Atlas*,
2021, <https://diabetesatlas.org/data/en/indicators/2/>.

82 There is, however, evidence that T1DM rates are also increasing
globally; Christopher C. Patterson et al., 'Worldwide estimates of
incidence, prevalence and mortality of type 1 diabetes in children
and adolescents: results from the International Diabetes Federation
Diabetes Atlas, 9th edition', *Diabetes Research and Clinical Practice* 157

(2019), https://doi.org/10.1016/j.diabres.2019.107842; Paula A. Diaz-Valencia et al., 'Global epidemiology of type 1 diabetes in young adults and adults: a systematic review', *BMC Public Health* 15 (2015), https://doi.org/10.1186/s12889-015-1591-y.

83 Spam is particularly associated with Hawaii, but it remains popular throughout the Pacific and South-East Asia, where it was widely introduced by the US military following the Second World War; Rachel Laudan, *Food of Paradise: Exploring Hawaii's Culinary Heritage* (Honolulu, HI: University of Hawai'i Press, 1996), pp. 66–9.

84 Karen E. Charlton et al., 'Fish, food security and health in Pacific Island countries and territories: a systematic literature review', *BMC Public Health* 16 (2016), https://doi.org/10.1186/s12889-016-2953-9; Katherine Sievert et al., 'Processed foods and nutrition transition in the pacific: regional trends, patterns, and food system drivers', *Nutrients* 11 (2019), https://doi.org/10.3390/nu11061328.

85 Kathryn Riley, 'A death and desperation in sanctions-hit Cuba', *The Guardian*, 18 July 2021, <https://www.theguardian.com/world/2021/jul/18/a-death-and-desperation-in-sanctions-hit-cuba>.

86 Timour Azhari, 'In Lebanon, diabetes medicine now costs more than monthly minimum wage', *Business Live*, 18 November 2021, <https://www.businesslive.co.za/bd/world/europe/2021-11-18-in-lebanon-diabetes-medicine-now-costs-more-than-monthly-minimum-wage/>.

Conclusion: Insulin for All?

1 Leslie Boehm and Gregory P. Marchildon, 'Canada needs a university-based, domestic vaccine-making capability', *Policy Options*, 22 March 2021, <https://policyoptions.irpp.org/magazines/march-2021/canada-needs-a-university-based-domestic-vaccine-making-capability/>.

2 For some particularly absurd examples, see the various YouTube channels operated by internet clown Nicholas Perry – better known by his stage name 'Nikocado Avocado'. For Christmas 2021, for example, he uploaded 'Diabetes . . . Mukbang', and spent the duration consuming particularly sweet and indulgent holiday foods – 'Mukbang' being an originally South Korean video format in which the host eats great amounts while talking. Diabetes features as a punch-line frequently in Perry's content; 'More Nikocado', 25 December 2021, 'Diabetes . . . Mukbang', *YouTube*, <https://youtu.be/DnR2azdoPZY>.

3 Kay Mellor (director), 'Episode #3.6', *The Syndicate* (2015), Rollem Productions.

4 Tommy Wirkola (director), *Hansel & Gretel: Witch Hunters* (2013), Paramount Pictures.

5 Matt Shakman (director), 'Frank's Pretty Woman', *It's Always Sunny in Philadelphia* (2011), 3 Arts Entertainment.

6 @Anniecoops ('Anne Cooper RN FQNI'), 'This is appalling and shows the ignorance of HCPs re Diabetes', *Twitter*, 13 August 2019, https://twitter.com/Anniecoops/status/1161173396262195200.

7 The quotations were taken from Jane K. Dickinson et al., 'Diabetes education as a career choice', *The Diabetes Educator* 41 (2015), p. 672.

8 @UrgoMedicalUK ('Urgo Medical UK'), 'The case for motivational interviewing', *Twitter*, 23 October 2019, https://twitter.com/UrgoMedicalUK/status/1187034276073824257.

9 @Dr_DianeJohnson ('Diane Johnson, PhD'), 'This is absolute proof that massive ignorance in some of the medical profession is alive & kicking when it comes to treating PWD', *Twitter*, 24 October 2019, https://twitter.com/Dr_DianeJohnson/status/1187394784333316098.

10 NHS England, *Language Matters* (2018), <https://www.england.nhs.uk/wp-content/uploads/2018/06/language-matters.pdf>.

11 @Kidfears99 ('Laura Marston'), 'Went to my PCP yesterday and had full bloodwork done', *Twitter*, 18 October 2019, https://twitter.com/Kidfears99/status/1185188858822086657.

12 @RenzaS ('Renza / Diabetogenic'), 'Okay diabetes healthcare professionals, listen up!', *Twitter*, 17 October 2019, https://twitter.com/RenzaS/status/1184677355442163714.

13 @Colonelblighty ('Guardian of the Glucose'), 'Australia is not alone in this sort of thing', *Twitter*, 17 October 2019, https://twitter.com/Colonelblighty/status/1184763278632898560; @Colonel Blighty ('Guardian of the Glucose'), 'In fairness it was from a few years ago', *Twitter*, 18 October 2019, https://twitter.com/Colonelblighty/status/1185302753536593920.

14 @RenzaS ('Renza / Diabetogenic'), 'It's like a fucking report card', *Twitter*, 17 October 2019, https://twitter.com/RenzaS/status/1184763597638864897.

15 One Twitter user, for example, reflected that for all their rhetoric,

she would 'love for an endocrinologist to trail me for a day and tell me exactly how much insulin to take, just to see if their guesses were any better than mine'; @MiriamETucker ('Miriam E. Tucker'), 'I would love for an endocrinologist to trail me for a day and tell me exactly how much insulin to take, just to see if their guesses were better than mine', *Twitter*, 14 October 2019, https://twitter.com/MiriamETucker/status/1183563971468681217.

16 Franz J. Ingelfinger, 'Arrogance', *New England Journal of Medicine* 303 (1980), p. 1509.

17 Annemarie Mol, *The Logic of Care: Health and the Problem of Patient Choice* (Abingdon: Routledge, 2008), p. 1.

18 Ibid., p. 13.

19 Ibid., p. 29.

20 Ibid., p. 22.

21 Ezekiel J. Emanuel and Linda L. Emanuel, 'Four models of the physician–patient relationship', *Journal of the American Medical Association* 267 (1992), p. 2222.

22 Ibid., p. 2222.

23 Stanley Joel Reiser, *Medicine and the Reign of Technology* (Cambridge: Cambridge University Press, 1978).

24 James A. Trostle, 'Medical compliance as an ideology', *Social Science and Medicine* 27 (1988), p. 1306.

25 Howie, Interview.

26 'Episode 165 – Diabetics doing sex work with Andi', *Diabetics Doing Things*, 23 March 2021, <https://diabeticsdoingthings.com/podcast?offset=1618466400130>.

27 For more on this idea, see Heather R. Walker and Michelle L. Litchman, 'Diabetes identity: a mechanism of social change', *Qualitative Health Research* 31 (2021), pp. 913–25.

28 @T1Bionic ('My Artificial Pancreas'), 'Patients have – medical domain expertise, device security expertise, tech expertise', *Twitter*, 25 September 2019, https://twitter.com/T1Bionic/status/1176638568569044994.

29 Sarah Zhang, 'People are clamouring to buy old insulin pumps', *The Atlantic*, 29 April 2019, <https://www.theatlantic.com/science/archive/2019/04/looping-created-insulin-pump-underground-market/588091/>.

30 For one personal account of 'DIY' looping using open-source projects, see Tim Omer, 'Health hacking', *Science Museum Blog*, 1 August 2016, <https://blog.sciencemuseum.org.uk/health-hacking/>.

31 In many respects, this is reminiscent of HIV/AIDS activism during the 1980s, when, frustrated by the perceived inadequacies of mainstream research, laypeople began to set up grassroots organizations to conduct their own community-based studies and directly challenge the traditional authorities on their own home turf. This is discussed further in Collins and Pinch, *Dr. Golem*, pp. 177–8.

32 Such events are often advertised via social media, for example this 'Build your own DIY artificial pancreas' workshop in London in early 2020; @Tims_Pants ('Tim S'), 'We're running a "Build your own DIY artificial pancreas" event in January at the #MicrosoftReactor site in London on 25th January, which makes it a no-cost event', *Twitter*, 10 October 2019, https://twitter.com/Tims_Pants/status/1182204961087725568.

33 The first of these was, unsurprisingly, Medtronic's. The company's MiniMed670G hybrid closed-loop system was initially approved in the United States in 2016; Aria Saunders et al., 'MiniMed 670G closed loop artificial pancreas system for the treatment of Type 1 diabetes mellitus: overview of its safety and efficacy', *Expert Review of Medical Devices* 16 (2019), pp. 845–53.

34 For example, Thomas S.J. Crabtree et al., 'Health-care professional opinions of DIY artificial pancreas systems in the UK', *Lancet Diabetes and Endocrinology* 8 (2020), pp. 186–7; Emma G. Wilmot and Thomas Danne, 'DIY artificial pancreas systems: the clinician perspective', *Lancet Diabetes and Endocrinology* 8 (2020), pp. 183–5.

35 Samantha Gottlieb, 'The FDA, patient empowerment, and the Type 1 diabetes communities in the era of digital health', *Platypus – The Castac Blog*, 23 April 2019, <http://blog.castac.org/2019/04/the-fda-patient-empowerment-and-the-type-1-diabetes-communities-in-the-era-of-digital-health/>.

36 The code necessary to force pumps to 'loop', for example, can be freely downloaded from the internet, and the developers provide comprehensive information and instructions for those wishing to implement the design. This can be found at *Openaps.org*, <https://openaps.org/>.

37 Rachel Brazil, 'Binning the sharps: the quest for oral insulin', *The*

Pharmaceutical Journal 303 (2019), https://doi.org/10.1211/PJ.20 19.20207045; Henry Anhalt and Nancy J.V. Bohannon, 'Insulin patch pumps: their development and future in closed-loop systems', *Diabetes Technology and Therapeutics* 12 (2010), pp. s51–s58; Elsemiek E.C. Engwerda et al., 'Needle-free jet injection of rapid-acting insulin improves postprandial glucose control in patients with diabetes', *Diabetes Care* 36 (2013), pp. 3436–41.

38 Avery Johnson, 'Insulin flop costs Pfizer $2.8 billion', *Wall Street Journal*, 19 October 2007, <https://www.wsj.com/articles/SB11926 9071993163273>.

39 'Diabetes inhaler rejected for NHS', *BBC*, 19 April 2006, <http://news.bbc.co.uk/1/hi/health/4919802.stm>.

40 Ann M. Thayer, 'No sugarcoating for insulin failures', *C&EN*, <https://cen.acs.org/articles/86/i19/Sugarcoating-Insulin-Failures.html>.

41 Subsequent long-term surveillance studies of those treated with Exubera have not been able to rule out a connection to the development of lung cancer; for example, Nicolle M. Gatto et al., 'Lung cancer-related mortality with inhaled insulin or a comparator: follow-up study of patients previously enrolled in Exubera controlled clinical trials (FUSE) final results', *Diabetes Care* 45 (2019), pp. 1708–15.

42 NICE, *Inhaled Insulin for the Treatment of Diabetes (Types 1 and 2)* (2006), p. 18.

43 For example, Clifford J. Bailey and Anthony H. Barnett, 'Why is Exubera being withdrawn?', *BMJ* 7630 (2007), p. 1156.

44 After the failure of Exubera, most major pharmaceutical companies withdrew funding for inhaled insulin research. Today, only one type – MannKind's Afrezza device, originally approved in the United States in 2014 – is available. Nonetheless, despite some cautious enthusiasm following clinical trials in recent years, it remains a highly niche piece of technology and is used by only a tiny minority of people with diabetes. For example, Alfonso Galderisi et al., 'Effect of Afrezza on glucose dynamics during HCL treatment', *Diabetes Care* 43 (2020), pp. 2146–52.

45 Alan C. Farney et al., 'Evolution of islet transplantation for the last 30 years', *Pancreas* 45 (2016), pp. 8–20.

46 Marco Farina et al., 'Cell encapsulation: overcoming barriers in cell transplantation in diabetes and beyond', *Advanced Drug Delivery*

Reviews 139 (2019), pp. 92–115; Andreas G. Tzakis et al., 'Pancreatic islet transplantation after upper abdominal exenteration and liver replacement', *The Lancet* 336 (1990), pp. 402–5.

47 'Lilly and Sigilon Therapeutics announce strategic collaboration to develop encapsulated cell therapies for the treatment of Type 1 diabetes', *Lilly Investors*, 4 April 2018, <https://investor.lilly.com/news-releases/news-release-details/lilly-and-sigilon-therapeutics-announce-strategic-collaboration>.

48 Mark A. Jarosinski et al., '"Smart" insulin-delivery technologies and intrinsic glucose-responsive insulin analogues', *Diabetologia* 64 (2021), pp. 1016–29; Thomas Hoeg-Jensen, 'Review: Glucose-sensitive insulin', *Molecular Metabolism* 46 (2021), https://doi.org/10.1016/j.molmet.2020.101107.

49 Merck actually ended up quietly dropping their GRI research programme following a disappointing clinical trial that ended in 2016; Dara Mohammadi, 'Towards a smarter insulin', *The Pharmaceutical Journal* 299 (2017), https://doi.org/10.1211/PJ.2017.20203828.

50 'Eli Lilly and Company acquired Glycostasis, Inc. from Codon Capital and Others', *MarketScreener*, 17 February 2016, <https://www.marketscreener.com/quote/stock/ELI-LILLY-AND-CO MPANY-13401/news/Eli-Lilly-and-Company-acquired-Glycostasis-Inc-from-Codon-Capital-and-others-38095174/>; 'Lilly announces acquisition of protomer technologies', *Lilly Investors*, 14 July 2021, <https://investor.lilly.com/news-releases/news-release-details/lilly-announces-acquisition-protomer-technologies>.

51 'Babbage: The shot of the century – 100 years of insulin', *The Economist*, 3 August 2021, <https://www.economist.com/podcasts/2021/08/03/the-shot-of-the-century-100-years-of-insulin>.

52 Charles Best, *Letter to Sir Henry Dale*, 22 February 1954, Ms. Coll. 235 (Feasby) Box 3, Folder 5, Feasby (William R.) Papers, Thomas Fisher Rare Book Library, University of Toronto.

53 For example, Rachel Delle Palme, 'Insulin patent sold for $1', *Banting House*, 14 December 2018, <https://bantinghousenhs.ca/20 18/12/14/insulin-patent-sold-for-1/>. I have been unable to find any direct source for this widely circulated quote, but it would certainly have fit with Banting's general attitude. For example, see Frederick

Banting, *Letter to Sir Robert Falconer*, 27 January 1923, A1982-0001, Box 62, Folder 1, University of Toronto Board of Governors Insulin Committee Archive, Archives and Records Management Services, University of Toronto.

54 'Eli Lilly', *Companies Market Cap*, <https://companiesmarketcap .com/eli-lilly/earnings/>.

55 Lauren Gambino, '"Insulin is our Oxygen": Bernie Sanders rides another campaign bus to Canada', *The Guardian*, 28 July 2019, <https://www.theguardian.com/us-news/2019/jul/28/bernie-sanders-americans-canada-insulin-bus-caravan>.

56 David Shultz, 'A black market for life-saving insulin thrives on social media', *Medium*, 15 January 2020, <https://onezero.medium .com/a-black-market-for-life-saving-insulin-thrives-on-social-media-5376ba3b3850>.

57 *Mutual Aid Diabetes*, <https://mutualaiddiabetes.com/>.

58 For example, @Kidfears99 ('Laura Marston'), 'Urgent need for Novolog and Levemir pens in Savannah, Ga.', *Twitter*, https://twit ter.com/Kidfears99/status/1448622839079256071.

59 *Open Insulin Foundation*, <https://openinsulin.org/>.

60 James S. Hirsch, *Cheating Destiny: Living with Diabetes, America's Biggest Epidemic* (Boston, MA: Houghton Mifflin, 2006), pp. 249–56.

61 Internal Revenue Service, 'Additional preventative care benefits permitted to be provided by a high deductible health plan under § 223', 17 July 2019, <https://www.irs.gov/pub/irs-drop/n-19-45.pdf>.

62 Annalisa Merelli, 'The insulin copay cap was a bad idea anyway', *Quartz*, 9 August 2022, <https://qz.com/the-insulin-copay-cap-was -a-bad-idea-anyway-1849385967>.

63 Michael Sainato, '"Seems like a scam": Americans with diabetes criticize Biden's insulin proposal', *The Guardian*, 4 January 2022, <https:// www.theguardian.com/us-news/2022/jan/04/insulin-copay-biden-build-back-better>.

64 Annalisa Van Den Bergh and Robin Cressman, 'To end the insulin crisis, we need to divest from diabetes nonprofits', *Jacobin*, 14 April 2022, <https://jacobin.com/2022/04/diabetes-nonprofits-insulin -manufacturers-jdrf-ada-cost>.

65 *Affordable Insulin Project*, <http://affordableinsulinproject.org/>.

66 *GetInsulin.org*, <https://getinsulin.org/>.

67 Christel Marchand Aprigliano, *Re: LD 673, An Act to Create the Insulin Safety Net Program*, 13 April 2021, Maine State Legislature, <https://legislature.maine.gov/backend/app/services/getDocument.aspx?doctype=test&documentId=148743>.

68 For example, 'How beyond Type 1 opposed the Insulin Safety Net Bill in Maine while we are in an insulin crisis', *Miss Diabetes*, <https://missdiabetes.com/how-beyond-type-1-is-blocking-access-to-insulin/>.

69 'Advocates celebrate passage of Alec Smith Bill in Minnesota', *T1International*, 15 April 2020, <https://www.t1international.com/blog/2020/04/15/alecs-insulin-bill-passes-minnesota/>. The bill is not, however, perfect by any stretch. In order to benefit, for example, individuals must provide identification, putting it out of reach of undocumented migrants.

70 California State Legislature, *California State Budget, 2022–2023* (2022), p. 82.

71 Eric Sagonowsky, 'Eli Lilly CEO's pay grows – again – to $23.7M after COVID antibody launch, new approvals', *Fierce Pharma*, 22 February 2021, <https://www.fiercepharma.com/pharma/after-navigating-pandemic-lilly-ceo-ricks-sees-2020-pay-jump-to-23-7m>.

72 Julia Kollewe, 'Pfizer accused of pandemic profiteering as profits double', *The Guardian*, 8 February 2022, <https://www.theguardian.com/business/2022/feb/08/pfizer-covid-vaccine-pill-profits-sales>.

73 Eric Sagonowsky, 'Pfizer, riding high on COVID-19 vaccine launch, pays CEO Bourla $21M for 2020, *Fierce Pharma*, 12 March 2021, <https://www.fiercepharma.com/pharma/pfizer-riding-high-covid-19-vaccine-launch-pays-ceo-bourla-21m-2020>.

74 'Coronavirus: WHO chief criticises "shocking" global vaccine divide', *BBC*, <https://www.bbc.co.uk/news/world-56698854>.

75 Stephanie Baker and Vernon Silver, 'Pandemic powers clash over access to shots', *Bloomberg*, 19 November 2021, <https://www.bloomberg.com/news/newsletters/2021-11-19/pandemic-powers-clash-over-access-to-shots>.

Selected Bibliography

General Histories of Diabetes and Insulin

For a condition so prevalent, politically relevant, and rich with meaning, there are fewer books on the general history of diabetes mellitus than might be expected, and many of them are now very dated. As far as broad survey works go, there are even fewer again. Tattersall's *The Biography* is, I think, by far the most accessible.

Hall, Kersten T., *Insulin – The Crooked Timber: A History from Thick Brown Muck to Wall Street Gold* (Oxford: Oxford University Press, 2022).

Hurley, Dan, *Diabetes Rising: How a Rare Disease Became a Modern Pandemic, and What to Do About It* (New York, NY: Kaplan, 2010).

Papaspyros, N.S., *The History of Diabetes Mellitus*, 2nd edn (Stuttgart: Georg Thieme Verlag, 1964).

Poulsen, Jacob E., *Features of the History of Diabetology* (Copenhagen: Munksgaard, 1982).

Tattersall, Robert, *Diabetes: The Biography* (Oxford: Oxford University Press, 2009).

Tattersall, Robert, *The Pissing Evil: A Comprehensive History of Diabetes Mellitus* (Fife: Swan & Horn, 2017).

The Discovery of Insulin

Despite being first published in 1982, Bliss' *The Discovery of Insulin* remains the essential work here. It is meticulously detailed and extremely thorough, and the story itself is timeless, though the revised 2007 anniversary edition is perhaps a little easy on companies like Eli Lilly. *Banting: A Biography* makes for a nice companion piece, but it is not vital.

Bliss, Michael, *The Discovery of Insulin, 25th Anniversary Edition* (Chicago, IL: University of Chicago Press, 2007).

Bliss, Michael, *Banting: A Biography* (Toronto, ON: Toronto University Press, 1992).

Cooper, Thea and Arthur Ainsberg, *Breakthrough: Elizabeth Hughes, the Discovery of Insulin, and the Making of a Medical Miracle* (New York, NY: St. Martin's Press, 2010).

Cox, Caroline, *The Fight to Survive: A Young Girl, Diabetes, and the Discovery of Insulin* (New York, NY: Kaplan Publishing, 2009).

Dworschak, Francisc Ion and Constantin Ionescu Tîrgovişte, *Paulescu and Collip: Insulin's Unsung Heroes* (Bucharest: Editura ILEX, 2008).

More Specific Aspects of the History of Diabetes

There are several books that examine certain topics within the history of diabetes in more depth. Of those listed here, I most enjoyed Feudtner's *Bittersweet*, which explores the experience of Elliott Joslin's patients at his clinic in Boston following the discovery of insulin. Tuchman's book on race is also excellent and very timely.

Feudtner, Chris, *Bittersweet: Diabetes, Insulin, and the Transformation of Illness* (Chapel Hill, NC: University of North Carolina Press, 2003).

Furdell, Elizabeth Lane, *Fatal Thirst: Diabetes in Britain until Insulin* (Leiden: Brill, 2009).

Hall, Stephen S., *Invisible Frontiers: The Race to Synthesize a Human Gene* (London: Sidgwick & Jackson, 1988).

Lawrence, Jane and Robert Tattersall (ed.), *Diabetes, Insulin and the Life of RD Lawrence* (London: Royal Society of Medicine Press, 2012), pp. 10–17.

Moore, Martin D., *Managing Diabetes, Managing Medicine: Chronic Disease and Clinical Bureaucracy in Post-War Britain* (Manchester: Manchester University Press, 2019).

Smith-Morris, Carolyn, *Diabetes among the Pima: Stories of Survival* (Tucson, AZ: University of Arizona Press, 2006).

Tuchman, Arleen Marcia, *Diabetes: A History of Race and Disease* (New Haven, CT: Yale University Press, 2020).

Chronic Disease and its Meanings

There is a lot more material on the experience of chronic disease more generally, both for the people affected and, in Bluebond-Langer's case, for their immediate families. I have also included Peitzman's book on kidney failure here, because the topic overlaps quite heavily with diabetes. Of all of these, I would recommend both Gawande's *Being Mortal* and Kleinman's *The Illness Narratives* as accessible introductions to subjectivities in health.

Anderson, Warwick and Ian R. Mackay, *Intolerant Bodies: A Short History of Autoimmunity* (Baltimore, MD: Johns Hopkins University Press, 2014).

Bluebond-Langner, Myra, *In the Shadow of Illness: Parents and Siblings of the Chronically Ill Child* (Princeton, NJ: Princeton University Press, 1996).

Brandt, Allan M. and Paul Rozin (eds), *Mortality and Health* (New York, NY: Routledge, 1997).

Frazer, Bianca C. and Heather R. Walker, *(Un)doing Diabetes: Representation, Disability, Culture* (London: Palgrave Macmillan, 2021).

Gawande, Atul, *Being Mortal: Illness, Medicine, and What Matters in the End* (London: Profile, 2014).

Kestenbaum, Victor, *The Humanity of the Ill: Phenomenological Perspectives* (Knoxville, TN: University of Tennessee Press, 1982).

Kleinman, Arthur, *The Illness Narratives: Suffering, Healing, and the Human Condition* (New York, NY: Basic Books, 1988).

Peitzman, Stephen J., *Dropsy, Dialysis, Transplant: A Short History of Failing Kidneys* (Baltimore, MD: Johns Hopkins University Press, 2007).

Rosenberg, Charles E., *Our Present Complaint: American Medicine, Then and Now* (Baltimore, MD: Johns Hopkins University Press, 2007).

Rosenberg, Charles E. and Janet Golden (eds), *Framing Disease: Studies in Cultural History* (New Brunswick, NJ: Rutgers University Press, 1997).

Sontag, Susan, *Illness as Metaphor & AIDS and its Metaphors* (London: Penguin, 1991).

Zola, Irving Kenneth, *Missing Pieces: A Chronicle of Living with a Disability* (Philadelphia, PA: Temple University Press, 1982).

Healthcare Professionals and Their Patients

There are many reasons why doctors might not get on with their patients. These books consider why. I recommend Collins and Pinch's *Dr. Golem* as an introduction. Foucault is not at all easy reading, but, for those interested in the topic, it is worth persevering.

Collins, Harry, *Tacit and Explicit Knowledge* (Chicago, IL: University of Chicago Press, 2010).

Collins, Harry and Trevor Pinch, *Dr. Golem: How to Think about Medicine* (Chicago, IL: University of Chicago Press, 2005).

Foucault, Michel, *The Birth of the Clinic: An Archaeology of Medical Perception [1963]*, trans. A.M. Sheridan (London: Tavistock, 1989).

Greene, Jeremy A., *Prescribing by Numbers: Drugs and the Definition of Disease* (Baltimore, MD: Johns Hopkins University Press, 2007).

Katz, Jay, *The Silent World of Doctor and Patient* (Baltimore, MD: Johns Hopkins University Press, 1984).

Reiser, Stanley Joel, *Medicine and the Reign of Technology* (Cambridge: Cambridge University Press, 1978).

Personal Experience

The testimony of people who actually use insulin is always valuable. While they used to be hard to come by as a genre, autobiographical

accounts are becoming more widespread. Corris' account is straightforward and frank, and I like the way Hirsch weaves snippets of the larger
history around his own story. Sparling's *Rage Bolus* is a little different
to the others as a collection of poems about life with diabetes. Some are
humorous, some are melancholy and others are angry, but they each
show deep insight. The website *Diabetes Stories* (http://diabetes-stories
.com) also contains a large selection of interview material from people
with the condition, some of them diagnosed in the early twentieth
century. This is no longer online, but can be accessed through the
internet archive at https://archive.org/.

Barnard, Celeste, *My Life as a Diabetic* (Bloomington, IN: Xlibris, 2018).
Cobb, Rachel Moser, *Diabetes – Fifty Years with the Needle: My Life with Insulin Therapy* (New York, NY: Vantage Press, 1996).
Corris, Peter, *Sweet & Sour: A Diabetic Life* (Lismore, NSW: Southern Cross University, 2000).
Dominick, Andie, *Needles: A Memoir* (London: Virago, 1998).
Hirsch, James S., *Cheating Destiny: Living with Diabetes, America's Biggest Epidemic* (Boston, MA: Houghton Mifflin, 2006).
Huzzey, Tony, *My Life with Diabetes: 61 Years of Carb Counting* (Rothersthorpe: Paragon, 2011).
Sparling, Kerri, *Rage Bolus & Other Poems* (Self-published, 2021).

Patient Guidebooks

Taking a look at the books physicians and other professionals write
for people with diabetes is an excellent way to explore their attitudes
towards the condition, their patients, and their position as a whole. I
have listed a few examples here, several of which were discussed in the
main text of this book, but there are countless others available.

Distiller, Larry A., *So you have Diabetes!* (Lancaster: MTP Press, 1980).
Hillson, Rowan, *Diabetes: A New Guide* (London: Optima, 1992).
Joslin, Elliott P., *A Diabetic Manual for the Mutual Use of Doctor and Patient* (Philadelphia, PA: Lea & Febiger, 1934), p. 152.

Kilo, Charles and Joseph R. Williamson, *Diabetes: The Facts That Let You Regain Control of Your Life* (New York, NY: John Wiley & Sons, 1987).

Lawrence, R.D., *The Diabetic Life*, various editions (London: J. & A. Churchill, 1925–1965).

Lawrence, R.D., *The Diabetic ABC*, various editions (London: H.K. Lewis, 1929–1967).

Scheiner, Gary, *Think Like a Pancreas: A Practical Guide to Managing Diabetes with Insulin* (Boston, MA: Da Capo, 2011).

Steel, Judith M. and Margaret Dunn, *Coping with Life on Insulin* (Edinburgh: W&R Chambers, 1987).

Index

Page numbers in *italic* refer to Figures